一步一步学

Spring Boot

微服务项目实战

（第2版）

黄文毅 著

清华大学出版社
北京

内 容 简 介

本书深入浅出地介绍了 Spring Boot 2.x 在 Web 应用开发中的各种技术。全书共 21 章，第 1 章由零开始，引导读者快速搭建 Spring Boot 开发环境，为之后 Spring Boot 的探险之旅奠定基础。第 2 章、第 3 章、第 10 章和第 13 章介绍了 Spring Boot 的数据访问应用，包括 Spring Boot 集成 Druid、Spring Data JPA 和 MyBatis 以及快速访问 MySQL 和 MongoDB 数据库。第 4~6 章，介绍了 Spring Boot 集成 Thymeleaf 模板引擎、事务及拦截器和监听器的应用。第 7~9 章介绍 Spring Boot 使用 Redis 缓存和 Quartz 定时器、集成 Log4J 日志框架和发送 Email 邮件。第 11~12 章介绍 Spring Boot 集成 ActiveMQ、异步调用和全局异常使用。第 14~15 章介绍 Spring Boot 应用监控和应用安全 Security。第 16~17 章介绍 Spring Boot 微服务在 Zookeeper 中的注册、Dubbo 的使用、多环境配置和使用以及在 Tomcat 中的部署应用。第 18~20 章主要探索 Spring Boot 的容器化、单元测试以及背后的原理和执行流程。第 21 章以秒杀系统为例，介绍 Spring Boot 在项目开发中的应用。

本书既介绍了 Spring Boot 背后的原理和执行流程，又突出了 Spring Boot 与各种热点技术的整合应用，适用于所有 Java 编程语言开发人员、Spring Boot 开发爱好者以及计算机专业的学生等，也很适合作为培训机构与大专院校的教学用书。

本书封面贴有清华大学出版社防伪标签，无标签者不得销售。
版权所有，侵权必究。侵权举报电话：010-62782989　13701121933

图书在版编目（CIP）数据

一步一步学 Spring Boot：微服务项目实战/黄文毅著. —2 版. —北京：清华大学出版社，2019.11
ISBN 978-7-302-54248-3

Ⅰ. ①一… Ⅱ. ①黄… Ⅲ. ①JAVA 语言—程序设计 Ⅳ. ①TP312.8

中国版本图书馆 CIP 数据核字（2019）第 258134 号

责任编辑：王金柱
封面设计：王　翔
责任校对：闫秀华
责任印制：丛怀宇

出版发行：清华大学出版社
　　　　　网　　址：http://www.tup.com.cn，http://www.wqbook.com
　　　　　地　　址：北京清华大学学研大厦 A 座　　邮　　编：100084
　　　　　社 总 机：010-62770175　　邮　　购：010-62786544
　　　　　投稿与读者服务：010-62776969，c-service@tup.tsinghua.edu.cn
　　　　　质 量 反 馈：010-62772015，zhiliang@tup.tsinghua.edu.cn
印 装 者：三河市君旺印务有限公司
经　　销：全国新华书店
开　　本：180mm×230mm　　印　　张：20.5　　字　　数：459 千字
版　　次：2018 年 8 月第 1 版　　2019 年 12 月第 2 版　　印　　次：2019 年 12 月第 1 次印刷
定　　价：69.00 元

产品编号：085519-01

前言

Spring Boot 是近几年非常流行的微服务框架，相对于其他开发框架，Spring Boot 不但使用更加简单，而且功能更加丰富，性能更加稳定和健壮。Spring Boot 是在 Spring 框架基础上创建的一个全新的框架，其设计目的是简化 Spring 应用的搭建和开发过程，使得开发人员不仅能提高开发速度和生产效率，而且能够增强系统的稳定性和扩展性。因此，Spring Boot 已经成为 Java 开发人员的必备技术。

本书是一本突出实战的 Spring Boot 2.x 专业著作，从零开始，引导读者循序渐进地掌握 Spring Boot 在 Web 开发中的各种技术，既注重基础、Spring Boot 背后的原理和执行流程，又突出 Spring Boot 与各种热点技术的整合应用，如 Spring Boot 集成 MyBatis、Active MQ、MongoDB、Redis 缓存、Zookeeper、Log4J 日志等，此外还介绍了微服务的多环境与部署、微服务的容器化、微服务测试等当前微服务开发中的必备技术。本书最后还给出了一个简易版本的高并发秒杀系统的开发，希望读者通过本项目的实践真正掌握真实项目的开发能力。

本书的大部分内容都来自笔者的开发实践，在讲解上力求通俗易懂，深入浅出，通过完整的项目实例，带领大家一步一步学习 Spring Boot。通过实战项目，学习 Spring Boot 的基础知识、Spring Boot 的开发技巧和 Spring Boot 的技术原理，最终达到融会贯通。

本书结构

本书共分 21 章，各章内容概要如下：

第 1 章介绍了开始学习 Spring Boot 之前的环境准备，如何 1 分钟快速搭建 Spring Boot 项目、Spring Boot 文件目录的简单介绍以及 Maven Helper 插件的安装和使用等。

第 2 章主要介绍如何安装和使用 MySQL、Spring Boot 集成 MySQL 数据库、Spring Boot 集成 Druid 以及通过实例讲解 Spring Boot 的具体运用。

第 3 章主要介绍 Spring Data JPA 核心接口及继承关系、在 Spring Boot 中集成 Spring Data JPA 以及如何通过 Spring Data JPA 实现增删改查及自定义查询等。

第 4 章主要介绍 Thymeleaf 模板引擎、Thymeleaf 模板引擎标签和函数、Spring Boot 中如何使用 Thymeleaf、集成测试以及 Rest Client 工具的使用。

第 5 章主要介绍 Spring 声明式事务、Spring 注解事务行为以及在 Spring Boot 中如何使用方法级别事务和类级别事务等。

第 6 章主要介绍如何在 Spring Boot 中使用过滤器 Filter 和监听器 Listener。

第 7 章主要介绍如何安装 Redis 缓存、Redis 缓存 5 种基本数据类型的增删改查、Spring Boot 中如何集成 Redis 缓存以及如何使用 Redis 缓存用户数据等。

第 8 章主要回顾 Log4j 基础知识、在 Spring Boot 中集成 Log4j、Log4j 在 Spring Boot 中的运用以及如何把日志打印到控制台和记录到日志文件中。

第 9 章主要介绍在 Spring Boot 中使用 XML 配置和 Java 注解两种方式定义和使用 Quartz 定时器，以及如何在 Spring Boot 中通过 JavaMailSender 接口给用户发送广告邮件等。

第 10 章主要介绍如何在 Spring Boot 中集成 MyBatis 框架以及通过 MyBatis 框架实现查询等功能，最后介绍如何使用 MyBatisCodeHelper 插件快速生成增删改查代码。

第 11 章主要介绍 ActiveMQ 的安装与使用、Spring Boot 集成 ActiveMQ、利用 ActiveMQ 实现异步发表微信说说以及 Spring Boot 异步调用@Async 等内容。

第 12 章主要介绍 Srping Boot 全局异常使用、自定义错误页面、全局异常类开发、Retry 重试机制的使用等内容。

第 13 章主要介绍如何安装和使用 MongoDB 数据库、NoSQL Manager for MongoDB 客户端的安装与使用以及在 Spring Boot 中集成 MongoDB 数据库开发简单的功能等内容。

第 14 章主要介绍 Spring Security 的基础知识、Spring Boot 如何集成 Spring Security、利用 Spring Security 实现授权登录，以及利用 Spring Boot 实现数据库数据授权登录等内容。

第 15 章主要介绍如何通过 Spring Boot 监控和管理应用、自定义监控端点以及自定义 HealthIndicator 等内容。

第 16 章主要介绍如何安装并运行 Zookeeper、Spring Boot 集成 Dubbo、spring-boot-book-v2 项目的服务拆分和实践、正式版 API 如何发布和服务注册等内容。

第 17 章主要介绍 Spring Boot 多环境配置及使用、Spring Boot 如何打成 war 包并部署到外部 Tomcat 服务器上等。

第 18 章主要介绍 Docker 技术以及如何将 Spring Boot 项目容器化。

第 19 章主要介绍 Spring Boot 单元测试、Mockito/PowerMockito 测试框架、H2 内存型数据库、REST API 测试以及性能测试等内容。

第 20 章主要回顾了 DemoApplication 入口类注解和 run 方法的原理，梳理了 Spring Boot 启动执行的流程和简单分析了 spring-boot-starter 的起步依赖原理，同时介绍了真实项目中的跨域问题、Spring Boot 优雅关闭以及如何将普通 Web 项目改造成 Spring Boot 项目等内容。

第 21 章主要介绍如何使用 Spring Boot 搭建一个高可用、高性能、高并发的秒杀系统。

学习本书的预备知识

Java 基础
读者需要掌握 J2SE 基础知识，这是最基本也是最重要的。

Java Web 开发技术
在项目实战中需要用到 Java Web 的相关技术，比如 Spring、HTML、Tomcat、MyBatis 等技术。

数据库基础
读者需要掌握主流数据库的基本知识，比如 MySQL 等。掌握基本的 SQL 语法以及常用数据库的安装。

本书使用的软件版本

本书项目实战开发环境为 Windows 10，开发工具使用 Intellij IDEA 2016.2，JDK 使用 1.8 版本，Tomcat 使用 8.0 版本，Spring Boot 使用最新版 2.1.6.RELEASE。

读者对象

本书适合所有 Java 编程语言开发人员，所有对 Spring Boot 感兴趣并希望使用 Spring Boot 开发框架进行开发的人员，缺少 Spring Boot 项目实战经验以及对 Spring Boot 内部原理感兴趣的开发人员。

源代码与视频教学下载

GitHub 源代码下载地址：

https://github.com/huangwenyi10/spring-boot-book-v2.git

扫描下述二维码可下载本书视频教学：

如果下载有问题，可发送电子邮件至 booksaga@126.com 获得帮助，邮件标题为"一步一步学 Spring Boot：微服务项目实战（第 2 版）下载资源"。

致谢

本书能够顺利出版，首先要感谢清华大学出版社王金柱编辑给笔者一次和大家分享技术、交流学习的机会，感谢王金柱编辑在本书出版过程的辛勤付出。

感谢厦门星耀蓝图科技有限公司，笔者的 Spring Boot 知识是在贵公司积累沉淀的，书中很多的知识点和项目实战经验都来源于贵公司，感谢公司总经理杨小雄、主管林良昆、架构师高志强、同事陈明元和林腾亚对笔者的关心和帮助。

感谢目前在职的上海美团科技有限公司，感谢公司主管王纪伟、导师叶永林、组长文慧、同事杨伟勤对笔者的栽培与帮助。

感谢笔者的家人和何庆华学长，他们对笔者生活的照顾使得笔者没有后顾之忧，全身心投入到本书的写作当中。

限于水平和写作时间，书中难免存在不足，欢迎广大读者批评指正。

黄文毅
2019 年 9 月 25 日

目　　录

第1章　第一个 Spring Boot 项目 ··· 1

1.1　Spring Boot 开发环境准备 ··· 1
1.1.1　安装 JDK ··· 1
1.1.2　安装 Intellij IDEA ··· 3
1.1.3　安装 Apache Maven ··· 3
1.2　一分钟快速搭建 Spring Boot 项目 ··· 5
1.2.1　使用 Spring Initializr 新建项目 ··· 5
1.2.2　测试 ··· 7
1.2.3　Spring Boot 三种启动方式 ··· 8
1.3　Spring Boot 文件目录介绍 ··· 9
1.3.1　工程目录 ··· 9
1.3.2　入口类 ··· 10
1.3.3　测试类 ··· 11
1.3.4　pom 文件 ··· 12
1.4　Spring Boot 2.x 新特性 ··· 14
1.4.1　配置变更 ··· 14
1.4.2　第三方类库升级 ··· 14
1.4.3　HTTP/2 支持 ··· 14
1.4.4　响应式 Spring 编程支持 ··· 14
1.4.5　其他新特性 ··· 15
1.5　Maven Helper 插件的安装和使用 ··· 15
1.5.1　Maven Helper 插件安装 ··· 15
1.5.2　Maven Helper 插件使用 ··· 16
1.6　思考题 ··· 16

第2章　集成 MySQL 数据库 ··· 18

2.1　MySQL 介绍与安装 ··· 18
2.1.1　MySQL 概述 ··· 18
2.1.2　MySQL 安装 ··· 18
2.2　集成 MySQL 数据库 ··· 19
2.2.1　引入依赖 ··· 20
2.2.2　添加数据库配置 ··· 20
2.2.3　设计表和实体 ··· 20
2.3　集成测试 ··· 22
2.3.1　测试用例开发 ··· 22
2.3.2　测试 ··· 23
2.3.3　Navicat for MySQL 客户端安装与使用 ··· 23

		2.3.4 Intellij IDEA 连接 MySQL	24
2.4	集成 Druid		25
	2.4.1	Druid 概述	25
	2.4.2	引入依赖	26
	2.4.3	Druid 配置	26
	2.4.4	开启监控功能	27
	2.4.5	测试	29
2.5	HikariCP 连接池		29
	2.5.1	HikariCP 概述	29
	2.5.2	HikariCP 的使用	30

第 3 章 集成 Spring Data JPA ... 32

3.1	Spring Data JPA 介绍		32
	3.1.1	Spring Data JPA 介绍	32
	3.1.2	核心接口 Repository	33
	3.1.3	接口继承关系图	34
3.2	集成 Spring Data JPA		34
	3.2.1	引入依赖	34
	3.2.2	继承 JpaRepository	35
	3.2.3	服务层类实现	37
	3.2.4	增删改查分页简单实现	39
	3.2.5	自定义查询方法	40
3.3	集成测试		42
	3.3.1	测试用例开发	42
	3.3.2	测试	43
3.4	思考题		43

第 4 章 Thymeleaf 模板引擎与集成测试 ... 44

4.1	Thymeleaf 模板引擎介绍		44
4.2	使用 Thymeleaf 模板引擎		45
	4.2.1	引入依赖	45
	4.2.2	控制层开发	46
	4.2.3	Thymeleaf 模板页面开发	47
4.3	集成测试		48
	4.3.1	测试	48
	4.3.2	REST Client 工具介绍	49
	4.3.3	使用 REST Client 测试	49

第 5 章 Spring Boot 事务支持 ... 50

5.1	Spring 事务介绍		50
	5.1.1	Spring 事务回顾	50
	5.1.2	Spring 声明式事务	51

		5.1.3 Spring 注解事务行为 ············· 51
5.2	Spring Boot 事务使用 ············· 53	
	5.2.1	Spring Boot 事务介绍 ············· 53
	5.2.2	类级别事务 ············· 53
	5.2.3	方法级别事务 ············· 54
	5.2.4	测试 ············· 55
5.3	思考题 ············· 56	

第 6 章 使用过滤器和监听器 ············· 57

- 6.1 Spring Boot 使用过滤器 Filter ············· 57
  - 6.1.1 过滤器 Filter 介绍 ············· 57
  - 6.1.2 过滤器 Filter 的使用 ············· 58
  - 6.1.3 测试 ············· 60
- 6.2 Spring Boot 使用监听器 Listener ············· 60
  - 6.2.1 监听器 Listener 介绍 ············· 60
  - 6.2.2 监听器 Listener 的使用 ············· 61
  - 6.2.3 测试 ············· 62

第 7 章 集成 Redis 缓存 ············· 63

- 7.1 Redis 缓存介绍 ············· 63
  - 7.1.1 Redis 概述 ············· 63
  - 7.1.2 Redis 服务器安装 ············· 63
  - 7.1.3 Redis 缓存测试 ············· 65
- 7.2 Spring Boot 集成 Redis 缓存 ············· 71
  - 7.2.1 Spring Boot 缓存支持 ············· 71
  - 7.2.2 引入依赖 ············· 71
  - 7.2.3 添加缓存配置 ············· 72
  - 7.2.4 测试用例开发 ············· 72
  - 7.2.5 测试 ············· 73
- 7.3 Redis 缓存在 Spring Boot 中的使用 ············· 74
  - 7.3.1 监听器 Listener 开发 ············· 74
  - 7.3.2 项目启动缓存数据 ············· 76
  - 7.3.3 更新缓存数据 ············· 76
  - 7.3.4 测试 ············· 77

第 8 章 集成 Log4J 日志 ············· 79

- 8.1 Log4J 概述 ············· 79
- 8.2 集成 Log4J2 ············· 81
  - 8.2.1 引入依赖 ············· 81
  - 8.2.2 添加 Log4J 配置 ············· 82
  - 8.2.3 创建 log4j2.xml 文件 ············· 82
- 8.3 使用 Log4J 记录日志 ············· 83

	8.3.1 打印到控制台	83
	8.3.2 记录到文件	84
	8.3.3 测试	86
8.4	思考题	87

第 9 章 Quartz 定时器和发送 Email 88

9.1	使用 Quartz 定时器	88
	9.1.1 Quartz 概述	88
	9.1.2 引入依赖	90
	9.1.3 定时器配置文件	90
	9.1.4 创建定时器类	92
	9.1.5 Spring Boot 扫描配置文件	94
	9.1.6 测试	94
9.2	Spring Boot 发送 Email	94
	9.2.1 Email 介绍	94
	9.2.2 引入依赖	95
	9.2.3 添加 Email 配置	95
	9.2.4 在定时器中发送邮件	96
	9.2.5 测试	99

第 10 章 集成 MyBatis 100

10.1	MyBatis 介绍	100
10.2	集成 MyBatis 的步骤	100
	10.2.1 引入依赖	100
	10.2.2 添加 MyBatis 配置	101
	10.2.3 Dao 层和 Mapper 文件开发	101
	10.2.4 测试	104

第 11 章 异步消息与异步调用 106

11.1	JMS 消息概述	106
11.2	Spring Boot 集成 ActiveMQ	107
	11.2.1 ActiveMQ 概述	107
	11.2.2 ActiveMQ 的安装	108
	11.2.3 引入依赖	109
	11.2.4 添加 ActiveMQ 配置	109
11.3	使用 ActiveMQ	110
	11.3.1 创建生产者	110
	11.3.2 创建消费者	113
	11.3.3 测试	114
11.4	Spring Boot 异步调用	118
	11.4.1 异步调用概述	118
	11.4.2 @Async 使用	118
	11.4.3 测试	119

第 12 章 全局异常处理与 Retry 重试 … 122

12.1 全局异常介绍 … 122
12.2 Spring Boot 全局异常使用 … 123
12.2.1 自定义错误页面 … 123
12.2.2 测试 … 124
12.2.3 全局异常类开发 … 124
12.2.4 测试 … 127
12.3 Retry 重试机制 … 127
12.3.1 Retry 重试概述 … 127
12.3.2 Retry 重试机制使用 … 128
12.3.3 测试 … 130

第 13 章 集成 MongoDB 数据库 … 131

13.1 MongoDB 数据库介绍 … 131
13.1.1 MongoDB 的安装 … 131
13.1.2 NoSQL Manager for MongoDB 客户端的使用 … 133
13.2 集成 MongoDB … 134
13.2.1 引入依赖 … 134
13.2.2 添加 MongoDB 配置 … 134
13.2.3 连接 MongoDB … 135
13.2.4 测试 … 137

第 14 章 集成 Spring Security … 138

14.1 Spring Security 概述 … 138
14.2 集成 Spring Security 的步骤 … 139
14.2.1 引入依赖 … 139
14.2.2 配置 Spring Security … 140
14.2.3 测试 … 142
14.2.4 数据库用户授权登录 … 142
14.2.5 测试 … 149

第 15 章 Spring Boot 应用监控 … 150

15.1 应用监控介绍 … 150
15.2 使用监控 … 151
15.2.1 引入依赖 … 151
15.2.2 添加配置 … 151
15.2.3 测试 … 152
15.3 自定义端点 … 155
15.3.1 自定义端点 EndPoint … 155
15.3.2 测试 … 156
15.3.3 自定义 HealthIndicator … 157
15.3.4 测试 … 160

15.4　保护 Actuator 端点 ·········· 161

第 16 章　集成 Dubbo 和 Zookeeper ·········· 163

16.1　Zookeeper 的介绍与安装 ·········· 163
　　16.1.1　Zookeeper 概述 ·········· 163
　　16.1.2　Zookeeper 的安装与启动 ·········· 164
16.2　Spring Boot 集成 Dubbo ·········· 165
　　16.2.1　Dubbo 概述 ·········· 165
　　16.2.2　服务与接口拆分思路 ·········· 167
　　16.2.3　服务与接口拆分实践 ·········· 167
　　16.2.4　正式版发布 ·········· 171
　　16.2.5　Service 服务端开发 ·········· 172
　　16.2.6　Service 服务注册 ·········· 173
　　16.2.7　Client 客户端开发 ·········· 174

第 17 章　多环境配置与部署 ·········· 175

17.1　多环境配置概述 ·········· 175
17.2　多环境配置的使用 ·········· 177
　　17.2.1　添加多个配置文件 ·········· 177
　　17.2.2　配置激活选项 ·········· 177
　　17.2.3　测试 ·········· 178
17.3　自定义属性与加载 ·········· 179
　　17.3.1　自定义属性 ·········· 179
　　17.3.2　参数间的引用 ·········· 181
　　17.3.3　使用随机数 ·········· 182
17.4　部署 ·········· 184
　　17.4.1　Spring Boot 内置 Tomcat ·········· 184
　　17.4.2　Intellij IDEA 配置 Tomcat ·········· 185
　　17.4.3　war 包部署 ·········· 187
　　17.4.4　测试 ·········· 187
17.5　热部署 ·········· 188
17.6　思考题 ·········· 189

第 18 章　微服务容器化 ·········· 191

18.1　Docker 概述 ·········· 191
　　18.1.1　Docker 的优势 ·········· 191
　　18.1.2　Docker 的基本概念 ·········· 193
　　18.1.3　Docker 架构 ·········· 194
　　18.1.4　Docker 的安装 ·········· 195
18.2　Docker 的常用命令 ·········· 198
18.3　制作与自动化构建镜像 ·········· 205
　　18.3.1　制作镜像 ·········· 205

	18.3.2 使用 Dockerfile 构建镜像	208
18.4	Spring Boot 集成 Docker	212

第 19 章 微服务测试 … 217

- 19.1 Spring Boot 单元测试 … 217
  - 19.1.1 关于测试 … 217
  - 19.1.2 微服务测试 … 218
- 19.2 Spring Boot 单元测试 … 220
- 19.3 Mockito/PowerMockito 测试框架 … 223
  - 19.3.1 Mockito 概述 … 223
  - 19.3.2 Mockito 简单实例 … 224
  - 19.3.3 PowerMock 概述 … 227
  - 19.3.4 PowerMockito 简单实例 … 228
- 19.4 H2 内存型数据库 … 231
  - 19.4.1 H2 概述 … 231
  - 19.4.2 Spring Boot 集成 H2 … 231
- 19.5 REST API 测试 … 235
  - 19.5.1 Postman 概述 … 235
  - 19.5.2 Postman 的简单使用 … 235
- 19.6 性能测试 … 238
  - 19.6.1 AB 概述 … 238
  - 19.6.2 AB 测试 … 239

第 20 章 Spring Boot 原理解析 … 241

- 20.1 回顾入口类 … 241
  - 20.1.1 DemoApplication 入口类 … 241
  - 20.1.2 @SpringBootApplication 的原理 … 242
  - 20.1.3 SpringApplication 的 run 方法 … 243
  - 20.1.4 SpringApplicationRunListeners 监听器 … 245
  - 20.1.5 ApplicationContextInitializer 接口 … 246
  - 20.1.6 ApplicationRunner 与 CommandLineRunner … 247
- 20.2 SpringApplication 执行流程 … 248
- 20.3 spring-boot-starter 原理 … 250
  - 20.3.1 自动配置条件依赖 … 250
  - 20.3.2 Bean 参数获取 … 255
  - 20.3.3 Bean 的发现与加载 … 256
  - 20.3.4 自定义 starter … 263
- 20.4 跨域访问 … 269
- 20.5 优雅关闭 … 270
  - 20.5.1 Java 优雅停机 … 270
  - 20.5.2 Spring Boot 优雅停机 … 273
- 20.6 将 SSM/Maven 项目改造为 Spring Boot 项目 … 276

20.6.1 创建 Maven 项目 276
20.6.2 第一种改造方法 278
20.6.3 第二种改造方法 279
20.7 思考题 281

第 21 章 实战高并发秒杀系统 283

21.1 秒杀系统业务 283
21.1.1 什么是秒杀 283
21.1.2 秒杀系统的工作流程 284
21.2 秒杀系统的简单实现 284
21.2.1 创建 Spring Boot 项目 284
21.2.2 库表设计与 Model 实体类 285
21.2.3 集成 MySQL 和 JPA 290
21.2.4 Service 服务层的设计与开发 292
21.2.5 Controller 控制层的设计与开发 295
21.2.6 前端页面的设计与开发 297
21.2.7 代码测试 300
21.2.8 总结 301
21.3 秒杀系统读优化 302
21.3.1 高并发读优化 302
21.3.2 使用 Redis 缓存 302
21.4 流量削峰 306
21.4.1 流量削峰的原因 306
21.4.2 集成 ActiveMQ 306
21.4.3 生产者开发 307
21.4.4 消费者开发 308
21.5 业务优化 310
21.5.1 答题/验证码 310
21.5.2 分时分段 311
21.5.3 禁用秒杀按钮 311
21.6 降级、限流、拒绝服务 311
21.6.1 降级 311
21.6.2 限流 312
21.6.3 拒绝服务 312
21.7 避免单点 312
21.8 总结 313

参考文献 314

第 1 章

第一个 Spring Boot 项目

Spring Boot 是目前流行的微服务框架,倡导"约定优先于配置",其设计目的是用来简化新 Spring 应用的初始化搭建以及开发过程。Spring Boot 提供了很多核心的功能,比如自动化配置、提供 starter 简化 Maven 配置、内嵌 Servlet 容器、应用监控等功能,让我们可以快速构建企业级应用程序。

本章主要介绍开始学习 Spring Boot 之前的环境准备,如何一分钟快速搭建 Spring Boot,Spring Boot 文件目录简单介绍,以及 Maven Helper 插件的安装和使用等内容。

1.1 Spring Boot 开发环境准备

在开始学习 Spring Boot 之前,我们需要准备好开发环境。本节将以 Windows 操作系统为例,介绍如何安装 JDK、Intellij IDEA 及 Apache Maven。如果你的电脑上已经安装了 JDK、Intellij IDEA 或者 Apache Maven,可以跳过本节内容。

1.1.1 安装 JDK

JDK(Java SE Development Kit)建议使用 1.8 及以上的版本,其官方下载路径为:http://www.oracle.com/technetwork/java/javase/downloads/jdk8-downloads-2133151.html。大家可以

根据自己 Windows 操作系统的配置选择合适的 JDK1.8 安装包，这里就不过多描述。

软件下载完成之后，双击下载软件，出现安装界面，如图 1-1 所示。一路单击【下一步】按钮即可完成安装。这里笔者把 JDK 安装在 C:\Program Files\Java\jdk1.8.0_77 下。

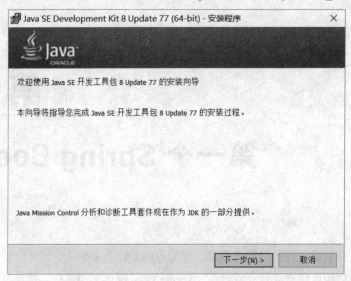

图 1-1　JDK 安装界面

安装完成后，需要配置环境变量 JAVA_HOME，具体步骤如下：

步骤 01　在电脑桌面上，右击【我的电脑】→【属性】→【高级系统设置】→【环境变量】→【系统变量(S)】→【新建】出现新建环境变量的窗口，如图 1-2 所示。

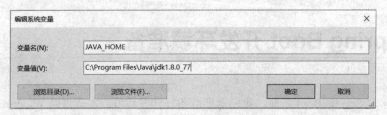

图 1-2　新建环境变量窗口

步骤 02　在【变量名】和【变量值】中分别输入 JAVA_HOME 和 C:\Program Files\Java\jdk1.8.0_77，单击【确定】按钮。

步骤 03　JAVA_HOME 配置好之后，将%JAVA_HOME%\bin 加入到【系统变量】的 path 中。完成后，打开命令行窗口，输入命令 java -version，如出现图 1-3 所示的提示，即表示安装成功。

图 1-3 安装成功命令行窗口

> 提示 JDK 安装路径最好不要出现中文，否则会出现意想不到的结果。

1.1.2 安装 Intellij IDEA

在 Intellij IDEA 的官方网站 http://www.jetbrains.com/idea/ 上可以免费下载 IDEA。下载完 IDEA 后，运行安装程序，按提示安装即可。本书使用 Intellij IDEA 2016.2 版本，当然大家也可以使用其他版本的 IDEA，只要版本不要过低即可。安装成功之后，打开软件界面如图 1-4 所示。

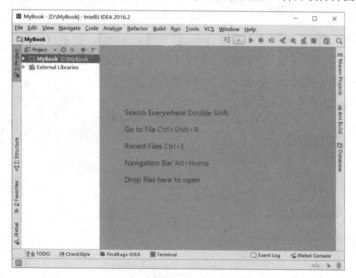

图 1-4 Intellij IDEA 软件窗口

1.1.3 安装 Apache Maven

Apache Maven 是目前流行的项目管理和构建自动化工具。虽然 IDEA 已经包含了 Maven 插件，但是笔者还是希望大家在工作中能够安装自己的 Maven 插件，方便以后项目配置需要。

大家可以通过 Maven 的官方网站 http://maven.apache.org/download.cgi 下载最新版的 Maven，本书 Maven 版本为 apache-maven-3.5.0。

下载完成后解压缩即可，例如，解压到 D：盘上，然后将 Maven 的安装路径 D:\apache-maven-3.5.0\bin 加入到 Window 的环境变量 path 中。安装完成后，在命令行窗口执行命令：mvn -v，如果输出如图 1-5 所示的页面，表示 Maven 安装成功。

图 1-5 Maven 安装成功命令行窗口

接下来，我们要在 Intellij IDEA 下配置 Maven，具体步骤如下：

步骤 01 在 Maven 安装目录，即 D:\apache-maven-3.5.0 下新建文件夹 repository，用来作为本地仓库。

步骤 02 在 Intellij IDEA 界面中，选择【File】→【Settings】，在出现的窗口中找到 Maven 选项，分别把【Maven home directory】【User settings file】【Local repository】，设置为我们自己 Maven 的相关目录，如图 1-6 所示。

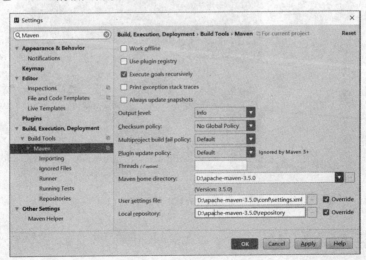

图 1-6 Maven 设置窗口

步骤 03 设置完成后，单击【Apply】→【OK】。至此，Maven 在 Intellij IDEA 的配置完成。

这里需要注意的是，之所以把 Maven 默认仓库（C:\${user.home}\.m2\respository）的路径改为我们自己的目录（D:\apache-maven-3.5.0\repository），是因为 respository 仓库到时候会存放很多的 jar 包，放在 C 盘影响电脑的性能，所以才会修改默认仓库的位置。

1.2 一分钟快速搭建 Spring Boot 项目

1.2.1 使用 Spring Initializr 新建项目

使用 Intellij IDEA 创建 Spring Boot 项目有多种，比如 Maven 和 Spring Initializr 方式。这里只介绍 Spring Initializr 这种方式，因为这种方式不但可为我们生成完整的目录结构，还可为我们生成一个默认的主程序，节省时间。我们的目的是掌握 Spring Boot 知识，而不是学一堆花样。具体步骤如下：

步骤01 在 Intellij IDEA 界面中，单击【File】→【New】→【Project】，在弹出的窗口中选择【Spring Initializr】选项，在【Project SDK】选择 JDK 的安装路径，如果没有则新建一个，单击【Next】按钮，如图 1-7 所示。

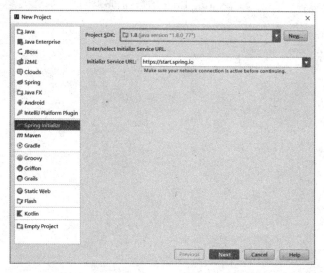

图 1-7 新建 Spring Boot 项目

步骤02 选择【Spring Boot Version】，这里按默认版本（本书使用的 Spring Boot 版本为 2.1.6）即可。勾选【web】→【Spring Web Starter】选项，然后单击【Next】按钮，如图 1-8 所示。

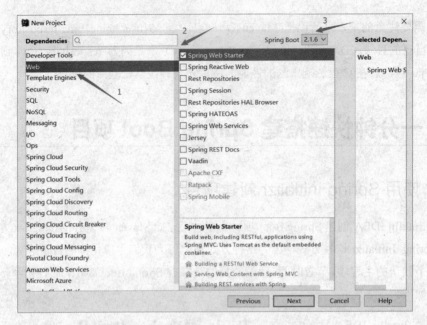

图 1-8 设置 Spring Boot 版本和组件

步骤 03 输入项目名称【spring-boot-book-v2】和项目存放目录,其他默认即可,然后单击【Finish】按钮。至此,一个完整的 Spring Boot 项目即创建完成,具体如图 1-9 所示。

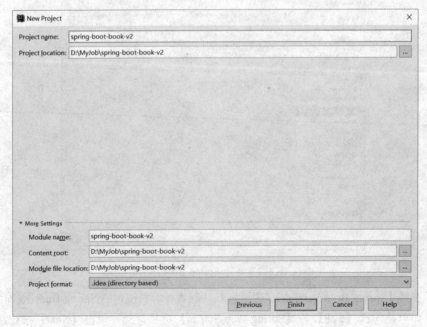

图 1-9 输入项目名称与存放目录

步骤04 如果开发工具弹出如图 1-10 所示的对话框,是提示你已经打开其他的项目,新创建的项目是否要在新的窗口打开。单击【New Window】表示在新的窗口打开项目,单击【This Window】表示在当前窗口打开项目,我们选择【New Window】按钮即可。

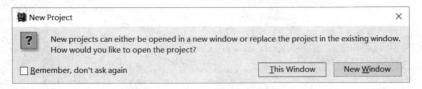

图 1-10　新窗口打开项目提示

步骤05 在 IDEA 开发工具上,找到刷新依赖的按钮(Reimport All Maven Projects),下载相关的依赖包,这时开发工具开始下载 Spring Boot 项目所需的依赖包,如图 1-11 所示。

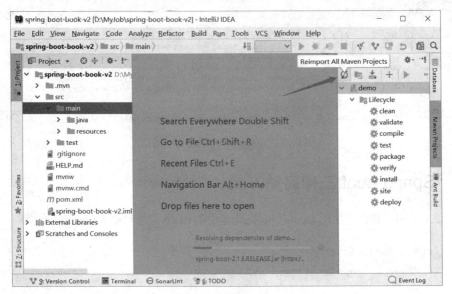

图 1-11　刷新依赖窗口

至此,Spring Boot 项目创建完成。

 下载 Spring Boot 依赖包是一个相对漫长的过程,可以去喝杯茶休息一会儿。

1.2.2　测试

Spring Boot 项目创建完成之后,找到入口类 DemoApplication 中的 main 方法并运行。当看到如图 1-12 所示,表示项目启动成功。同时还可以看出项目启动的端口(8080)及启动时间。

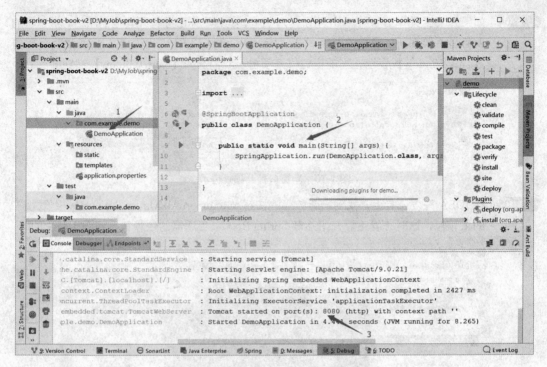

图 1-12 Spring Boot 项目启动成功

1.2.3 Spring Boot 三种启动方式

在 1.2.1 节中，我们是使用 DemoApplication 中的 main 方法启动 Spring Boot，即：

```
@SpringBootApplication
public class DemoApplication {

    public static void main(String[] args) {
        SpringApplication.run(DemoApplication.class, args);
    }
}
```

除此之外，还有两种方式可以启动 Spring Boot 项目：

（1）在 Spring Boot 应用的根目录下运行 mvn spring-boot:run

```
### 在命令行窗口执行 mvn spring-boot:run 命令
D:\MyJob\ssm-to-springboot>mvn spring-boot:run
```

（2）使用 java -jar 命令

```
### java -jar xxx.jar 命令
D:\MyJob\ssm-to-springboot\target>java -jar demo-0.0.1-SNAPSHOT.jar
```

1.3 Spring Boot 文件目录介绍

创建 Spring Boot 项目后，会产生一个工程目录，该工程目录存放了工程项目的各种文件，对于 Spring Boot 开发人员来说，了解该工程目录非常必要。

1.3.1 工程目录

Spring Boot 的工程目录，如图 1-13 所示。

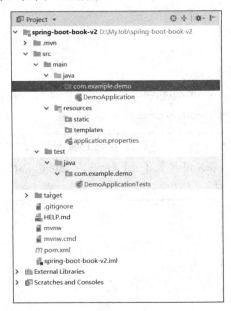

图 1-13　Spring Boot 工程目录

说明如下：

- /src/main/java：目录下放置所有的Java文件（源代码文件）。
- /src/main/resources：用于存放所有的资源文件，包括静态资源文件、配置文件、页面文件等。

- /src/main/resources/static：用于存放各类静态资源。
- /src/main/resources/templates：用于存放模板文件，如Thymeleaf（这个技术不懂不用着急，以后会介绍）模板文件。
- /src/main/resources/application.properties：配置文件，这个文件非常重要。Spring Boot 默认支持两种配置文件类型（.properties和.yml）。
- /src/test/java：放置单元测试类java代码。
- /target：放置编译后的.class文件和配置文件等。

Spring Boot 将很多配置文件进行了统一管理，且配置了默认值。Spring Boot 会自动在 /src/main/resources 目录下找 application.properties 或者 application.yml 配置文件，找到后将运用此配置文件中的配置，否则使用默认配置。这两种类型的配置文件有其一即可，也可以两者并存。两者的区别如下：

```
application.properties:
server.port = 8080
application.yml:
server:
      port:8080
```

application.properties 和 application.yml 配置文件的区别主要是书写格式不同，另外，application.yml 格式不支持@PropertySource 注解导入配置。

.properties 配置文件的优先级高于.yml。在.properties 文件中配置了 server.port = 8080，同时，在.yml 配置了 server.port = 8090，Spring Boot 将使用.properties 中的 8080 端口。

1.3.2　入口类

入口类的代码很简单，代码如下：

```
package com.example.demo;
import org.springframework.boot.SpringApplication;
import org.springframework.boot.autoconfigure.SpringBootApplication;
//入口类
@SpringBootApplication
public class DemoApplication {
```

```
    public static void main(String[] args) {
        SpringApplication.run(DemoApplication.class, args);
    }

}
```

- @SpringBootApplication：是一个组合注解，包含@EnableAutoConfiguration、@ComponentScan和@SpringBootConfiguration三个注解，是项目启动注解。如果我们使用这三个注解，项目依旧可以启动起来，只是过于烦琐。此时，可用@SpringBootApplication简化。
- @SpringApplication.run：应用程序开始运行的方法。

 入口类需要放置在包的最外层，以便能够扫描到所有子包中的类。

1.3.3 测试类

Spring Boot 的测试类主要放置在/src/test/java 目录下面。项目创建完成后，Spring Boot 会自动为我们生成测试类 DemoApplicationTests.java，测试类的代码如下：

```
package com.example.demo;
import org.junit.Test;
import org.junit.runner.RunWith;
import org.springframework.boot.test.context.SpringBootTest;
import org.springframework.test.context.junit4.SpringRunner;

@RunWith(SpringRunner.class)
@SpringBootTest
public class DemoApplicationTests {

    @Test
    public void contextLoads() {
    }

}
```

- @RunWith(SpringRunner.class)：@RunWith(Parameterized.class)是一个参数化运行器，可用于配合@Parameters使用JUnit的参数化功能。查源代码可知，SpringRunner类是继承的SpringJUnit4ClassRunner类，此处表明使用SpringJUnit4ClassRunner执行器，此执行器集成了Spring的一些功能。如果只是简单的JUnit单元测试，该注解可以去掉。
- @SpringBootTest：此注解能够测试我们的SpringApplication，因为Spring Boot程序的入口是SpringApplication，基本上所有配置都会通过入口类去加载，而该注解可以引用入口类的配置。
- @Test：JUnit单元测试的注解，注解在方法上表示一个测试方法。

当我们右键执行 DemoApplicationTests.java 中的 contextLoads 方法的时候，大家可以看到控制台打印的信息和执行入口类中的 SpringApplication.run()方法，打印的信息是一致的。由此便知，@SpringBootTest 是引入了入口类的配置。

1.3.4　pom 文件

Spring Boot 项目下的 pom.xml 文件主要用来存放依赖信息，具体代码如下（部分代码已省略）：

```xml
<?xml version="1.0" encoding="UTF-8"?>
<project xmlns="http://maven.apache.org/POM/4.0.0"
xmlns:xsi="http://www.w3.org/2001/XMLSchema-instance"
     xsi:schemaLocation="http://maven.apache.org/POM/4.0.0 http://maven.apache.org/xsd/maven-4.0.0.xsd">
    <modelVersion>4.0.0</modelVersion>
    <parent>
        <groupId>org.springframework.boot</groupId>
        <artifactId>spring-boot-starter-parent</artifactId>
        <version>2.1.6.RELEASE</version>
        <relativePath/> <!-- lookup parent from repository -->
    </parent>
    <groupId>com.example</groupId>
    <artifactId>demo</artifactId>
    <version>0.0.1-SNAPSHOT</version>
    <name>demo</name>
    <description>Demo project for Spring Boot</description>

    <properties>
```

```xml
        <java.version>1.8</java.version>
    </properties>

    <dependencies>
        <dependency>
            <groupId>org.springframework.boot</groupId>
            <artifactId>spring-boot-starter-web</artifactId>
        </dependency>

        <dependency>
            <groupId>org.springframework.boot</groupId>
            <artifactId>spring-boot-starter-test</artifactId>
            <scope>test</scope>
        </dependency>
    </dependencies>

    <build>
        <plugins>
            <plugin>
                <groupId>org.springframework.boot</groupId>
                <artifactId>spring-boot-maven-plugin</artifactId>
            </plugin>
        </plugins>
    </build>

</project>
```

- **spring-boot-starter-parent**：是一个特殊的starter，它用来提供相关的Maven默认依赖，使用它之后，常用的包依赖可以省去version标签。
- **spring-boot-starter-web**：只要将其加入到项目的maven依赖中，我们就会得到一个可执行的Web应用。该依赖中包含许多常用的依赖包，比如spring-web、spring-webmvc等。这样，我们不需要做任何Web配置，便能获得相关Web服务。
- **spring-boot-starter-test**：这个依赖和测试相关，只要引入它，就会把所有与测试相关的包全部引入。
- **spring-boot-maven-plugin**：这是一个Maven插件，能够以Maven的方式为应用提供Spring Boot的支持，即为Spring Boot应用提供执行Maven操作的可能，并能够将Spring Boot应用打包为可执行的jar或war文件。

1.4　Spring Boot 2.x 新特性

1.4.1　配置变更

在 2.x 中废除了一些 1.x 中的配置，并增加了许多新配置，详细请查看以下链接中的变更表格：

https://github.com/spring-projects/spring-boot/wiki/Spring-Boot-2.0-Configuration-Changelog

1.4.2　第三方类库升级

Spring Boot 2.x 对第三方类库升级了所有能升级的稳定版本，一些值得关注的类库升级如下：

（1）Spring Framework 5+
（2）Tomcat 8.5+
（3）Flyway 5+
（4）Hibernate 5.2+
（5）Thymeleaf 3+

Spring Boot 2.x 至少需要 JDK 8 的支持，2.x 应用了 JDK 8 的许多高级新特性，所以当应用要升级到 2.0 版本时，先确认应用必须兼容 JDK 8。

另外，2.x 开始了对 JDK 9 的支持。

1.4.3　HTTP/2 支持

提供对 HTTP/2 的支持，如 Tomcat、Undertow、Jetty，该功能依赖具体选择的应用服务器和应用环境。

1.4.4　响应式 Spring 编程支持

2.x 通过启动器和自动配置全面支持 Spring 的响应式编程，响应式编程是完全异步和非阻塞的，它是基于事件驱动模型，而不是传统的线程模型。就连 Spring Boot 内部也对一些功能点进行了有必要的响应式升级，最值得注意的是对内嵌式容器的支持。

对响应式编程支持又包括以下几个技术模块：

（1）Spring WebFlux & WebFlux.fn 支持。
（2）响应式 Spring Data 支持。
（3）响应式 Spring Security 支持。
（4）内嵌式的 Netty 服务器支持。

1.4.5 其他新特性

除了前面几节列出的变化外，还包括其他新特性：

（1）全面重写了 Spring Boot 的 Gradle 插件，并且最小支持 Gradle 4+，以便提供一些重要的特性提升。

（2）2.x 开始提供对 Kotlin 1.2 的支持，并且提供了一个 runApplication 函数来运行 Spring Boot 应用。

（3）在 2.x 中，对执行器端点进行了改进，所有的 HTTP 执行端点都暴露在/actuator 路径下，并对 JSON 结果集也做了改善。

（4）2.x 默认使用 HikariCP 连接池。

（5）提供了一个 spring-boot-starter-json 启动器对 JSON 读写的支持。

（6）2.x 提供了一个 spring-boot-starter-quartz 启动器对定时任务框架 Quartz 的支持。

（7）所有支持的容器都支持过滤器的初始化。

1.5 Maven Helper 插件的安装和使用

Maven Helper 是一款可以方便查看 Maven 依赖树的插件，可以在 Intellij IDEA 上安装使用它。它支持多种视图来查看 Maven 依赖，同时可以帮助我们分析 pom 文件中的依赖是否存在冲突，方便快速定位错误。

1.5.1 Maven Helper 插件安装

安装 Maven Helper 插件很简单，在菜单栏选择【File】→【Settings】→【Plugins】，在搜索框中输入【Maven Helper】，然后单击【Install】安装即可。安装成功之后重启 Intellij IDEA 即可使用。

1.5.2　Maven Helper 插件使用

插件安装成功之后打开 pom 文件，我们看到除了 Text 视图外，多了一个 Dependency Analyzer 视图，如图 1-14 所示。

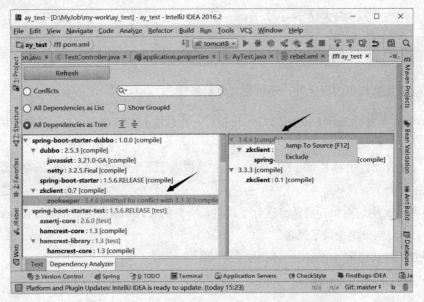

图 1-14　Maven Helper 插件

- Conflicts：查看所有有冲突的依赖包，如果存在冲突，会显示红色，同时会在右侧显示视图。我们可以单击有冲突的包，【右键】→【Exclude】来排除冲突，通过【右键】→【Jump To Source[F12]】可以跳转到源代码。
- All Dependencies as List：以列表方式显示所有的依赖包。
- All Dependencies as Tree：以树形方式显示所有的依赖包。

1.6　思考题

1. 什么是 Spring Boot？

答：Spring Boot 是 Spring 开源组织下的子项目，是 Spring 组件的一站式解决方案。Spring Boot 主要是简化了使用 Spring 的难度，节省了繁重的配置，提供了各种启动器，使开发者能快速上手。

2. Spring Boot 的优点主要有哪些？

答：Spring Boot 的优点非常多，如独立运行、简化配置、自动配置、无代码生成和 XML 配置、应用监控、健康检查、上手容易，等等。

3. 运行 Spring Boot 有哪几种方式？

答：可以使用 3 种方式运行 Spring Boot：（1）运行 main 方法启动 Spring Boot；（2）在 Spring Boot 应用的根目录下运行 mvn spring-boot:run；（3）使用 java -jar 命令。

4. Spring Boot 2.X 有什么新特性？

答：Spring Boot 2.X 的新特性主要有配置变更、JDK 版本升级、第三方类库升级、响应式 Spring 编程支持、HTTP/2 支持、配置属性绑定等。

5. 创建 Spring Boot 项目最简单的方法是什么？

答：Spring Initializr http://start.spring.io/ 是引导 Spring Boot 项目的绝佳工具。

第 2 章

集成 MySQL 数据库

本章将介绍如何安装和使用 MySQL、Spring Boot 集成 MySQL 数据库、Spring Boot 集成 Druid 以及通过实例讲解 Spring Boot 的具体运用。

2.1 MySQL 介绍与安装

数据库类型有很多，比如像 MySQL、Oracle 这样的关系型数据库，又比如像 MongoDB、NoSQL 这样的非关系型数据库。本节主要讲解目前项目中运用广泛的关系型数据库 MySQL。

2.1.1 MySQL 概述

MySQL 是目前项目中运用广泛的关系型数据库，无论是普通公司还是互联网公司都运用甚广。MySQL 所使用的 SQL 语言是用于访问数据库的最常用的标准化语言。MySQL 软件由于其体积小、速度快、总体拥有成本低，尤其是开放源代码这一特点，一般中小型网站的开发都选择 MySQL 作为网站数据库。

2.1.2 MySQL 安装

MySQL 的安装很简单，安装方式也有多种。大家可以到 MySQL 的官方网站

https://dev.mysql.com/downloads/mysql/ 去下载 MySQL 安装软件，并按照提示一步一步安装即可。如果你的电脑上已经安装了 MySQL，可略过此节。本书使用的 MySQL 版本为 5.7.17。

安装完成之后，我们需要检验 MySQL 安装是否成功。具体步骤如下：

步骤 01　打开命令行窗口，进入 MySQL 安装目录，笔者的 MySQL 安装目录是 C:\Program Files\MySQL\MySQL Server 5.7\bin。

步骤 02　在命令行窗口中输入命令 mysql -uroot -p 和密码登录 MySQL，然后再输入命令 status，出现如图 2-1 所示的信息，表示安装成功。

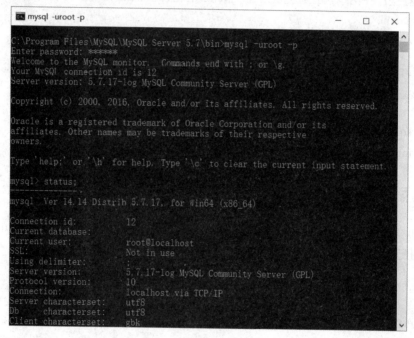

图 2-1　MySQL 安装状态

2.2　集成 MySQL 数据库

由于 Spring Boot 包含一个功能强大的资源库，所以 Spring Boot 集成 MySQL 非常简单。借助于 Spring Boot 框架，可以不用编写原始的访问数据库的代码，也不用调用 JDBC（Java Data Base Connectivity）或者连接池等诸如此类的被称为底层的代码，就可以在更高级的层次上访问数据库。

2.2.1 引入依赖

集成 MySQL 数据库之前,需要在项目中的 pom 文件中添加 MySQL 所需的依赖,具体代码如下:

```xml
<dependency>
    <groupId>mysql</groupId>
    <artifactId>mysql-connector-java</artifactId>
</dependency>
<dependency>
    <groupId>org.springframework.boot</groupId>
    <artifactId>spring-boot-starter-jdbc</artifactId>
</dependency>
```

- mysql-connector-java:MySQL 连接 Java 的驱动程序。
- spring-boot-starter-jdbc:支持通过 JDBC 连接数据库。

2.2.2 添加数据库配置

在 pom 文件中引入 MySQL 所需的 Maven 依赖之后,需要在 application.properties 文件中添加如下的配置信息:

```
### MySQL 连接信息
spring.datasource.url=jdbc:mysql://127.0.0.1:3306/test?serverTimezone=UTC
###用户名
spring.datasource.username=root
###密码
spring.datasource.password=123456
###驱动
spring.datasource.driver-class-name=com.mysql.jdbc.Driver
```

2.2.3 设计表和实体

配置信息添加完成之后,下面我们在 MySQL 数据库中创建一张表。MySQL 安装成功之后,默认有一个 test 数据库,这里在 test 数据库里新建表 ay_user。具体创建表的 SQL 语句如下:

```
-- ----------------------------
-- 用户表
-- Table structure for ay_user
-- ----------------------------
DROP TABLE IF EXISTS `ay_user`;
CREATE TABLE `ay_user` (
  `id` varchar(32) NOT NULL COMMENT '主键',
  `name` varchar(10) DEFAULT NULL COMMENT '用户名',
  `password` varchar(32) DEFAULT NULL COMMENT '密码'
);
```

数据库表 ay_user 字段很简单，主键（id）、用户名（namc）和密码（password）。ay_user 表创建好之后，我们往数据库表 ay_user 插入两条数据，具体插入数据的 SQL 语句如下：

```
INSERT INTO `ay_user` (`id`, `name`, `password`) VALUES ('1', '阿毅', '123456');
INSERT INTO `ay_user` (`id`, `name`, `password`) VALUES ('2', '阿兰', '123456');
```

除了使用 SQL 语句插入之外，还可以使用 Navicat for MySQL 客户端插入数据，下一节会详细介绍。数据插入成功之后，可在 MySQL 客户端查询到这两条数据，具体如图 2-2 和图 2-3 所示。

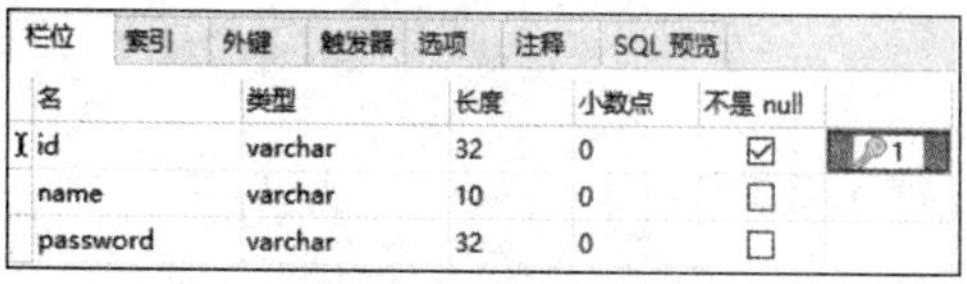

名	类型	长度	小数点	不是 null	
id	varchar	32	0	☑	1
name	varchar	10	0	☐	
password	varchar	32	0	☐	

图 2-2　创建表 ay_user

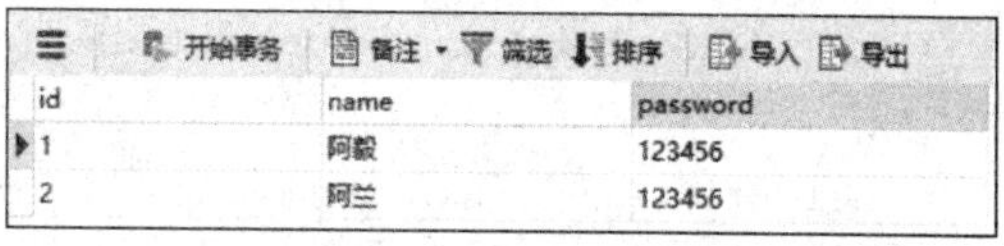

id	name	password
1	阿毅	123456
2	阿兰	123456

图 2-3　插入 2 条数据

表和数据准备好之后，我们在项目的目录下（/src/main/java/com.example.demo.model）新建实体类，具体代码如下：

```
/**
 * 描述：用户表
 * @Author 阿毅
 * @date   2017/10/8
 */
public class AyUser {

    //主键
    private String id;
```

```
    //用户名
    private String name;
    //密码
    private String password;

    //此处省略 set、get 方法
}
```

至此，数据库表、数据、实体已经全部准备好了，接下来我们开发测试用例进行测试。

2.3 集成测试

2.3.1 测试用例开发

我们在项目的单元测试类 DemoApplicationTests.java 中添加如下代码：

```
@Resource
    private JdbcTemplate jdbcTemplate;
    /**
     * MySQL 集成 Spring Boot 简单测试
     */
    @Test
    public void mySqlTest(){
        String sql = "select id,name,password from ay_user";
        List<AyUser> userList =
        (List<AyUser>) jdbcTemplate.query(sql, new RowMapper<AyUser>(){
            @Override
            public AyUser mapRow(ResultSet rs, int rowNum) throws SQLException {
                AyUser user = new AyUser();
                user.setId(rs.getString("id"));
                user.setName(rs.getString("name"));
                user.setPassword(rs.getString("password"));
                return user;
            }
        });
        System.out.println("查询成功: ");
        for(AyUser user:userList){
```

```
            System.out.println("【id】: " + user.getId() + "; 【name】: " + user.getName());
        }
    }
```

- JdbcTemplate：这是一个通过JDBC连接数据库的工具类。其中上一节我们引入依赖spring-boot-starter-jdbc中包含的spring-jdbc包下，可以通过这个工具类对数据库进行增删改查等操作。
- @Resource：自动注入，通过这个注解在项目启动之后，Spring Boot会帮助我们实例化一个JdbcTemplate对象，省去初始化工作。
- query()方法：JdbcTemplate对象中的查询方法，通过传入SQL语句和RowMapper对象，可以查询出数据库中的数据。
- RowMapper对象：RowMapper对象可以将查询出的每一行数据封装成用户定义的类，在上面的代码中，通过调用RowMapper中的方法mapRow，可将数据库中的每一行数据封装成AyUser对象，并返回去。

 SQL语句要么全部大写，要么全部小写，请不要大小写混用。

2.3.2 测试

单元测试方法开发完成之后，双击方法 mySqlTest，用右键执行一下单元测试，这时可以在控制台上看到打印信息，具体如下：

查询成功：

【id】: 1; 【name】: 阿毅

【id】: 2; 【name】: 阿兰

至此，Spring Boot 集成 MySQL 数据库大功告成。这一节的内容简单但是非常重要，之后的章节都是在本节的基础上进行开发的。

2.3.3 Navicat for MySQL 客户端安装与使用

Navicat for MySQL 是连接 MySQL 数据库的客户端工具，通过使用该客户端工具，可以方便我们对数据库进行操作，比如创建数据库表、添加数据等。如果大家已经安装了其他的 MySQL 客户端，可以略过本节。

Navicat for MySQL 的安装也非常简单，可以到网上下载安装即可。安装完成之后，我们打开软件，如图 2-4 所示。

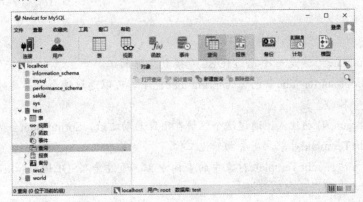

图 2-4　创建表 ay_user

可以通过【查询】→【新建查询】，在弹出的窗口中编写相关的查询语句来查询数据。当然还有很多的操作，大家可以自己去使用和掌握它，这里就不一一描述了。

2.3.4　Intellij IDEA 连接 MySQL

除了通过 Navicat for MySQL 客户端连接数据库之外，如果你不喜欢在自己的电脑上安装一堆软件的话，还可以通过 Intellij IDEA 来连接 MySQL 数据库。具体步骤如下：

步骤 01　在 Intellij IDEA 中，单击右侧的【Database】→【New（加号）】→【Data Source】→【MySQL】，在弹出的窗口中输入主机、用户名、密码、端口等信息，如图 2-5 所示。

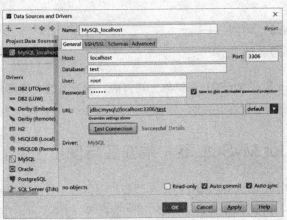

图 2-5　Intellij IDEA 连接 MySQL

步骤02 单击【Test Connection】测试是否连接成功，然后单击【Apply】→【OK】。

步骤03 连接成功之后，可以看到如图 2-6 所示的界面，【双击】数据库表 ay_user，可以看到如图 2-7 所示的界面。

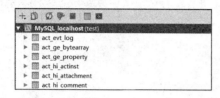

图 2-6 连接 MySQL 成功界面　　　　　图 2-7 双击表 ay_user 界面

成功连接 MySQL 数据库之后，可以在图 2-6 中看到停止数据库、刷新数据库、命令行窗口等按钮，通过这些按钮可以停止和刷新数据库，或者打开命令行窗口编写 SQL 语句。在图 2-7 中，可以查询某张表的数据，单击 + 号、- 号按钮进行数据的添加和删除，还可以在 Filter criteria 输入框中编写过滤条件，搜索出所需要的数据。比如在 Filter criteria 输入框中输入 id = '1' and name = '阿毅' 或者 name like '%兰%' and id = '2'，查询结果如图 2-8 和图 2-9 所示。

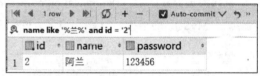

图 2-8 连接 MySQL 成功界面（一）　　　　　图 2-9 连接 MySQL 成功界面（二）

2.4 集成 Druid

2.4.1 Druid 概述

Druid 是阿里巴巴开源项目中的一个数据库连接池。Druid 是一个 JDBC 组件，它包括三部分：

（1）DruidDriver：代理 Driver，能够提供基于 Filter-Chain 模式的插件体系。

（2）DruidDataSource：高效可管理的数据库连接池。

（3）SQLParser：它支持所有 JDBC 兼容的数据库，包括 Oracle、MySQL、SQL Server 等。

Durid 在监控、可扩展、稳定性和性能方面具有明显的优势。通过其提供的监控功能，可以实现观察数据库连接池和 SQL 查询的工作情况。使用 Druid 连接池，可以提高数据库的访问性能。

2.4.2 引入依赖

我们在项目中的 pom 文件继续添加 Durid 的依赖，具体代码如下：

```
<dependency>
    <groupId>com.alibaba</groupId>
    <artifactId>druid</artifactId>
    <version>1.1.18</version>
</dependency>
```

在这里，笔者使用的是最新版本 1.1.4，添加完依赖之后，Intellij IDEA 会自动帮助我们下载依赖包，我们只要刷新下依赖即可。

2.4.3 Druid 配置

依赖添加完成之后，在 application.properties 配置文件中继续添加 Druid 配置，之前我们已经添加了 MySQL 的连接 URL、用户名、密码等配置，application.properties 完整代码如下：

```
### MySQL 连接信息
spring.datasource.url=jdbc:mysql://127.0.0.1:3306/test?serverTimezone=UTC
spring.datasource.username=root
spring.datasource.password=123456
spring.datasource.driver-class-name=com.mysql.jdbc.Driver
###  数据源类别
spring.datasource.type=com.alibaba.druid.pool.DruidDataSource
### 初始化大小，最小，最大
spring.datasource.initialSize=5
spring.datasource.minIdle=5
spring.datasource.maxActive=20
### 配置获取连接等待超时的时间，单位是毫秒
spring.datasource.maxWait=60000
### 配置间隔多久才进行一次检测，检测需要关闭的空闲连接，单位是毫秒
spring.datasource.timeBetweenEvictionRunsMillis=60000
### 配置一个连接在池中最小生存的时间，单位是毫秒
spring.datasource.minEvictableIdleTimeMillis=300000
spring.datasource.validationQuery=SELECT 1 FROM DUAL
spring.datasource.testWhileIdle=true
spring.datasource.testOnBorrow=false
```

```
spring.datasource.testOnReturn=false
### 打开PSCache，并且指定每个连接上PSCache的大小
spring.datasource.poolPreparedStatements=true
spring.datasource.maxPoolPreparedStatementPerConnectionSize=20
### 配置监控统计拦截的filters，去掉后监控界面SQL无法统计，'wall'用于防火墙
spring.datasource.filters=stat,wall,log4j
### 通过connectProperties属性来打开mergeSql功能；慢SQL记录
spring.datasource.connectionProperties=druid.stat.mergeSql=true;druid.stat.slowSqlMillis=5000
### 合并多个DruidDataSource的监控数据
#spring.datasource.useGlobalDataSourceStat=true
```

上面每一个配置的含义都有相关的注释，这里就不再过多介绍。这里要注意的是，在.properties 配置文件中，# 字符是注释符号。

2.4.4　开启监控功能

开启监控功能的方式有多种：

（1）使用原生的 Servlet 和 Filter 方式，然后通过@ServletComponentScan 启动扫描包的方式进行处理。

（2）使用代码注册 Servlet 和 Filter 的方式处理。

这里我们选择 Spring Boot 推荐的第二种方式实现。我们在项目 Java 目录下（/src/main/java/com.example.demo.filter）新建一个配置类 DruidConfiguration.java。具体代码如下：

```
@Configuration
public class DruidConfiguration {

    @Bean
    public ServletRegistrationBean druidStatViewServle(){
        //ServletRegistrationBean提供类的进行注册
        ServletRegistrationBean servletRegistrationBean
                = new ServletRegistrationBean(new StatViewServlet(),
"/druid/*");
        //添加初始化参数：initParams
        //白名单：
        servletRegistrationBean.addInitParameter("allow","127.0.0.1");
```

```java
            //IP黑名单 (存在共同时，deny优先于allow)
            // 如果满足deny的话提示:Sorry, you are not permitted to view this page.
            servletRegistrationBean.addInitParameter("deny","192.168.1.73");
            //登录查看账号和密码的信息
            servletRegistrationBean.addInitParameter("loginUsername","admin");
            servletRegistrationBean.addInitParameter("loginPassword","123456");
            //是否能够重置数据
            servletRegistrationBean.addInitParameter("resetEnable","false");
            return servletRegistrationBean;
        }

        @Bean
        public FilterRegistrationBean druidStatFilter(){
            FilterRegistrationBean filterRegistrationBean
                    = new FilterRegistrationBean(new WebStatFilter());
            //添加过滤规则
            filterRegistrationBean.addUrlPatterns("/*");
            //添加需要忽略的格式信息
            filterRegistrationBean.addInitParameter("exclusions",
                    "*.js,*.gif,*.jpg,*.png,*.css,*.ico,/druid/*");
            return filterRegistrationBean;
        }
    }
```

- @Configuration：Spring中有很多的XML配置文件，文件中会配置很多的bean。在类上添加@Configuration注解，可以理解为该类变成了一个XML配置文件。
- @Bean：等同于XML配置文件中的<bean>配置。Spring Boot会把加上该注解的方法的返回值装载进Spring IoC容器，方法的名称对应<bean>标签的id属性值。具体代码如下：

```java
@Bean
public FilterRegistrationBean druidStatFilter(){
    FilterRegistrationBean filterRegistrationBean
                = new FilterRegistrationBean(new WebStatFilter());
    return filterRegistrationBean;
}
```

等同于：

```
<bean id="druidStatFilter"
class="org.springframework.boot.web.servlet.ServletRegistrationBean">
</bean>
```

- 类ServletRegistrationBean和FilterRegistrationBean：在DruidConfiguration.java这个配置文件中我们配置了两个类：druidStatViewServlet和druidStatFilter，并且通过注册类ServletRegistrationBean和FilterRegistrationBean实现Servlet和Filter类的注册。

在druidStatViewServlet类中，设定了访问数据库的白名单、黑名单、登录用户名和密码等信息。在druidStatFilter类中，设定了过滤的规则和需要忽略的格式。至此，配置类开发完成。

2.4.5 测试

在 DruidConfiguration.java 类开发完成之后，重新启动项目，然后通过访问网址 http://localhost:8080/druid/index.html 打开监控的登录界面，如图 2-10 所示。在登录界面中输入用户名：admin，密码：123456，即可登录成功，如图 2-11 所示。

图 2-10 Druid 监控登录界面

图 2-11 Druid 登录成功界面

在 Druid 的监控界面中，我们可以对数据源、SQL、Web 应用等进行监控。

2.5 HikariCP 连接池

2.5.1 HikariCP 概述

HikariCP 是数据库连接池，而且号称史上最快的。在 Spring Boot 2.0 版本中，由于 HikariCP 提供了卓越的性能，默认数据库池技术已从 Tomcat Pool 切换到 HikariCP。spring-boot-starter-jdbc 和 spring-boot-starter-data-jpa 默认解析 HikariCP 依赖，spring.datasource.type 属性将 HikariDataSource 作为默认值。

HikariCP 为什么这么快，主要原因有以下几点：

（1）代码量非常小

对于连接池来讲，代码越少，占用 CPU 和内存越少，Bug 出现概率也就越小，执行率高。这就是为什么 HikariCP 受欢迎的原因之一。

（2）稳定性，可靠性强

HikariCP 经受了市场的考验。

（3）速度快

优化并精简了字节码，可以更好地并发集合类实现 ConcurrentBag，使用 FastList 替代 ArrayList 等。

2.5.2 HikariCP 的使用

如果我们使用的是 Spring Boot 2.0 或者之后的版本，不需要单独在 pom.xml 文件中引入 HikariCP 依赖，因为默认情况下 spring-boot-starter-jdbc 或者 spring-boot-starter-data-jpa 会依赖进来。

对于 Hikari 连接池的配置，可通过使用 spring.datasource.type 并在 application.properties 文件中为其分配连接池实现的完全限定名称来启用它，如下所示：

```
spring.datasource.type = com.zaxxer.hikari.HikariDataSource
```

如果使用的是 Spring Boot 2.0 及以上版本，Spring Boot 会使用 HikariDataSource 作为默认选择，不需要配置上面的行。

要配置 Hikari 特定的连接池设置，Spring Boot 提供了 spring.datasource.hikari.* 在 application.properties 文件中使用的前缀。我们将在这里讨论一些常用的配置。

（1）connectionTimeout

connectionTimeout 是客户端等待连接池连接的最大毫秒数，需要将其配置如下：

```
spring.datasource.hikari.connection-timeout=20000
```

（2）minimumIdle

minimumIdle 是 HikariCP 在连接池中维护的最小空闲连接数，它配置如下：

```
spring.datasource.hikari.minimum-idle=5
```

（3）maximumPoolSize

maximumPoolSize 配置最大池大小，它配置如下：

```
spring.datasource.hikari.maximum-pool-size=12
```

（4）idleTimeout

idleTimeout 是允许连接在连接池中空闲的最长时间（以毫秒为单位），它配置如下：

```
spring.datasource.hikari.idle-timeout=300000
```

（5）maxLifetime

maxLifetime 是池中连接关闭后的最长生命周期（以毫秒为单位），它配置如下：

```
spring.datasource.hikari.max-lifetime=1200000
```

（6）autoCommit

autoCommit 配置从池中返回的连接的默认自动提交行为，默认值为 true：

```
spring.datasource.hikari.auto-commit=true
```

完整的配置示例如下：

```
spring.datasource.url=jdbc:mysql://localhost:3306/concretepage
spring.datasource.username=root
spring.datasource.password=cp

#Spring Boot 2.0 includes HikariDataSource by default
#spring.datasource.type = com.zaxxer.hikari.HikariDataSource

spring.datasource.hikari.connection-timeout=20000
spring.datasource.hikari.minimum-idle=5
spring.datasource.hikari.maximum-pool-size=12
spring.datasource.hikari.idle-timeout=300000
spring.datasource.hikari.max-lifetime=1200000
spring.datasource.hikari.auto-commit=true

spring.jpa.properties.hibernate.dialect=org.hibernate.dialect.MySQLDialect
spring.jpa.properties.hibernate.id.new_generator_mappings=false
spring.jpa.properties.hibernate.format_sql=true
```

本节只是简单地介绍了 HikariCP 的基础知识，其具体使用与 Druid 连接池类似。

第 3 章
集成 Spring Data JPA

本章主要介绍 Spring Data JPA 核心接口及继承关系，在 Spring Boot 中集成 Spring Data JPA，以及如何通过 Spring Data JPA 实现增删改查及自定义查询等内容。

3.1 Spring Data JPA 介绍

本节主要介绍 Spring Data JPA 是什么、Spring Data JPA 核心接口 Repository、核心接口间的继承关系图等内容。

3.1.1 Spring Data JPA 介绍

JPA（Java Persistence API）是 Sun 公司官方提出的 Java 持久化规范。所谓规范是指只定义标准规则，不提供实现，而 JPA 的主要实现有 Hibernate、EclipseLink、OpenJPA 等。JPA 是一套规范，不是一套产品，Hibernate 是一套产品，如果这些产品实现了 JPA 规范，那么我们可以把它们叫作 JPA 的实现产品。

Spring Data JPA 是 Spring Data 的一个子项目，它通过提供基于 JPA 的 Respository，极大地减少了 JPA 作为数据访问方案的代码量。通过 Spring Data JPA 框架，开发者可以省略实现持久层业务逻辑的工作，唯一要做的，就只是声明持久层的接口，其他都交给 Spring Data JPA 来帮你完成。

3.1.2 核心接口 Repository

Spring Data JPA 的最顶层接口是 Repository，该接口是所有 Repository 类的父类，具体代码如下：

```
package org.springframework.data.repository;
import java.io.Serializable;
public interface Repository<T, ID extends Serializable> {

}
```

Repository 类下没有任何的接口，只是一个空类。Repository 接口的子类有 CrudRepository、PagingAndSortingRepository、JpaRepository 等。其中 CrudRepository 类提供了基本的增删改查等接口，PagingAndSortingRepository 类提供了基本的分页和排序等接口，而 JpaRepository 是 CrudRepository 和 PagingAndSortingRepository 的子类，继承了它们的所有接口。在真实的项目当中，我们都是通过实现 JpaRepository 或者其子类进行基本的数据库操作，JpaRepository 的具体代码如下：

```
@NoRepositoryBean
public interface JpaRepository extends PagingAndSortingRepository<T, ID> {
    List<T> findAll();
    List<T> findAll(Sort var1);
    List<T> findAll(Iterable<ID> var1);
    <S extends T> List<S> save(Iterable<S> var1);
    void flush();
    <S extends T> S saveAndFlush(S var1);
    void deleteInBatch(Iterable<T> var1);
    void deleteAllInBatch();
    T getOne(ID var1);
    <S extends T> List<S> findAll(Example<S> var1);
    <S extends T> List<S> findAll(Example<S> var1, Sort var2);
}
```

- @NoRepositoryBean：使用该注解标明，此接口不是一个Repository Bean。

3.1.3 接口继承关系图

Repository 接口间的继承关系如图 3-1 所示。通过该继承图,可以清楚地知道接口间的集成关系。在项目中,一般都是实现 JapRepository 类,加上自己定义的业务方法,来完成我们的业务开发。

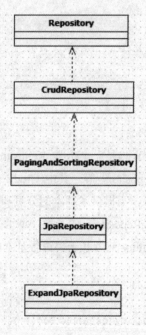

图 3-1　Repository 接口间集成关系

3.2　集成 Spring Data JPA

本节主要介绍如何在 Spring Boot 中集成 Spring Data JPA,服务层类开发,如何通过 Spring Data JPA 实现基本增删改查功能,以及自定义查询方法等内容。

3.2.1　引入依赖

在 Spring Boot 中集成 Spring Data JPA,首先需要在 pom.xml 文件中引入所需的依赖,具体代码如下:

```
<dependency>
    <groupId>org.springframework.boot</groupId>
    <artifactId>spring-boot-starter-data-jpa</artifactId>
</dependency>
```

在之前的章节中，我们已经在开发工具中安装好 Maven Helper 插件，所以可以通过该插件查看目前引入的所有依赖，具体如图 3-2 所示。

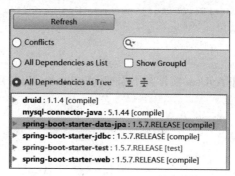

图 3-2　Maven Helper 查看 pom 依赖包

3.2.2　继承 JpaRepository

在 pom.xml 文件中引入依赖之后，我们在目录/src/main/java/com.example.demo.repository 下开发一个 AyUserRepository 类，如图 3-3 所示。

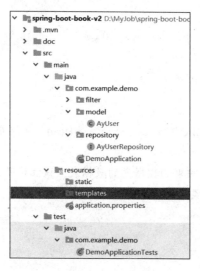

图 3-3　spring-boot-book-v2 项目目录

具体代码如下:

```java
/**
 * 描述: 用户Repository
 * @author 阿毅
 * @date   2017/10/14
 */
public interface AyUserRepository extends JpaRepository<AyUser,String>{

}
```

与此同时，我们需要在 AyUser 实体类下添加@Entity 和@Id 注解，具体代码如下：

```java
/**
 * 描述: 用户表
 * @Author 阿毅
 * @date   2017/10/8
 */
@Entity
@Table(name = "ay_user")
public class AyUser {
    //主键
    @Id
    private String id;
    //用户名
    private String name;
    //密码
    private String password;
}
```

- @Entity：每个持久化POJO类都是一个实体Bean，通过在类的定义中使用@Entity注解来进行声明。
- @Table：声明此对象映射到数据库的数据表。该注释不是必需的，如果没有则系统使用默认值（实体的短类名）。
- @Id：指定表的主键。

3.2.3 服务层类实现

我们在 spring-boot-book-v2 项目下继续开发服务层接口类和实现类：AyUserService 和 AyUserServiceImpl 类，它们分别存放在目录 /src/main/java/com.example.demo.service 和 /src/main/java/com.example.demo.service.impl 下。具体代码如下：

```
/**
 * 描述：用户服务层接口
 * @author 阿毅
 * @date   2017/10/14
 */
public interface AyUserService {
AyUser findById(String id);
    List<AyUser> findAll();
    AyUser save(AyUser ayUser);
    void delete(String id);
}
```

接口类 AyUserService 定义了 4 个接口，findById 和 findAll 用来查询单个和所有数据，Delete 用来删除数据，Save 同时具备保存和更新数据的功能。接口实现类 AyUserServiceImpl 代码如下：

```
/**
 * 描述：用户服务层实现类
 * @author 阿毅
 * @date   2017/10/14
 */
@Service
public class AyUserServiceImpl implements AyUserService{

    @Resource
    private AyUserRepository ayUserRepository;

    @Override
    public AyUser findById(String id){
        return ayUserRepository.findById(id).get();
    }
```

```java
    @Override
    public List<AyUser> findAll() {
        return ayUserRepository.findAll();
    }

    @Override
    public AyUser save(AyUser ayUser) {
        return ayUserRepository.save(ayUser);
    }

    @Override
    public void delete(String id) {
        ayUserRepository.deleteById(id);
    }
}
```

- @Service：Spring Boot会自动扫描到@Component注解的类，并把这些类纳入进Spring容器中管理。也可以用@Component注解，只是@Service注解更能表明该类是服务层类。
- @Component：泛指组件，当组件不好归类的时候，可以使用这个注解进行标注。
- @Repository：持久层组件，用于标注数据访问组件，即DAO组件。
- @Resource：这个注解属于J2EE的，默认按照名称进行装配，名称可以通过name属性进行指定。如果没有指定name属性，当注解写在字段上时，默认取字段名进行查找。如果注解写在setter方法上默认取属性名进行装配。当找不到与名称匹配的bean时才按照类型进行装配。但是需要注意的是，如果name属性一旦指定，就只会按照名称进行装配。具体代码如下：

```java
@Resource(name = "ayUserRepository")
private AyUserRepository ayUserRepository;
```

- @Autowired：这个注解是属于Spring的，默认按类型装配。默认情况下要求依赖对象必须存在，如果要允许null值，可以设置它的required属性为false，如：@Autowired(required=false)，如果想使用名称装配可以结合@Qualifier注解进行使用。具体代码如下：

```java
@Autowired
@Qualifier("ayUserRepository")
private AyUserRepository ayUserRepository;
```

3.2.4 增删改查分页简单实现

上一节，我们已经在服务层类 AyUserService 中开发完增删改查方法，这一节，我们将继续在类中添加分页接口，具体代码如下：

```
/**
 * 描述：用户服务层接口
 * @author 阿毅
 * @date    2017/10/14
 */
public interface AyUserService {
AyUser findById(String id);
    List<AyUser> findAll();
    AyUser save(AyUser ayUser);
    void delete(String id);
    //分页
        Page<AyUser> findAll(Pageable pageable);
}
```

- Pageable：这是一个分页接口，查询时只需要传入一个 Pageable接口的实现类，指定pageNumber和PageSize即可。pageNumber为第几页，而PageSize为每页大小。
- Page：分页查询结果会封装在该类中，Page接口实现Slice接口，通过查看其源代码可知。通过调用getTotalPages和 getContent等方法，可以方便获得总页数和查询的记录。Page接口和Slice接口的源代码如下：

```
public interface Page<T> extends Slice<T> {
    int getTotalPages();
    long getTotalElements();
    <S> Page<S> map(Converter<? super T, ? extends S> var1);
}

public interface Slice<T> extends Iterable<T> {
    int getNumber();
    int getSize();
    int getNumberOfElements();
    List<T> getContent();
    boolean hasContent();
```

```
    Sort getSort();
    boolean isFirst();
    boolean isLast();
    boolean hasNext();
    boolean hasPrevious();
    Pageable nextPageable();
    Pageable previousPageable();
    <S> Slice<S> map(Converter<? super T, ? extends S> var1);
}
```

分页方法定义好之后，在类 AyUserServiceImpl 中实现该方法，具体代码如下：

```
@Override
public Page<AyUser> findAll(Pageable pageable) {
    return ayUserRepository.findAll(pageable);
}
```

3.2.5 自定义查询方法

我们除了使用 JpaRepository 接口提供的增删改查分页等方法之外，还可以自定义查询方法。下面在 AyUserRepository 类中添加几个自定义查询方法，具体代码如下：

```
/**
 * 描述：用户 Repository
 * @author 阿毅
 * @date   2017/10/14
 */
public interface AyUserRepository extends JpaRepository<AyUser,String>{

    /**
     * 描述：通过名字相等查询，参数为 name
     * 相当于：select u from ay_user u where u.name = ?1
     */
    List<AyUser> findByName(String name);

    /**
     * 描述：通过名字 like 查询，参数为 name
     * 相当于：select u from ay_user u where u.name like ?1
     */
```

```
    List<AyUser> findByNameLike(String name);

    /**
     * 描述:通过主键 ID 集合查询,参数为 ID 集合
     * 相当于: select u from ay_user u where id in(?,?,?)
     * @param ids
     */
    List<AyUser> findByIdIn(Collection<String> ids);
}
```

在 AyUserRepository 中,我们自定义了 3 个查询的方法。从代码可以看出,Spring Data JPA 为我们约定了一系列的规范,只要按照规范编写代码,Spring Data JPA 就会根据代码翻译成相关的 SQL 语句,进行数据库查询。比如,可以使用 findBy、Like、In 等关键字,其中 findBy 可以用 read、readBy、query、queryBy、get、getBy 来代替。关于查询关键字的更多内容,可以到官方网站(https://docs.spring.io/spring-data/data-jpa/docs/current/reference/html/)查看,里面有详细的内容介绍,这里就不一一列举了。

AyUserRepository 类中自定义查询方法开发完成之后,可分别在类 AyUserService 和类 AyUserServiceImpl 中调用它们。

在 AyUserService 类中继续添加这 3 个方法,具体代码如下:

```
List<AyUser> findByName(String name);
List<AyUser> findByNameLike(String name);
List<AyUser> findByIdIn(Collection<String> ids);
```

在 AyUserServiceImpl 类中添加这 3 个方法,具体代码如下:

```
@Override
public List<AyUser> findByName(String name){
    return ayUserRepository.findByName(name);
}
@Override
public List<AyUser> findByNameLike(String name){
    return ayUserRepository.findByNameLike(name);
}
@Override
public List<AyUser> findByIdIn(Collection<String> ids){
    return ayUserRepository.findByIdIn(ids);
}
```

@Override 注解不可去掉，它可以帮助校验接口方法是否被误改。

3.3 集成测试

3.3.1 测试用例开发

我们在测试类 MySpringBootApplicationTests 中添加如下代码：

```
@Resource
private AyUserService ayUserService;

@Test
public void testRepository(){
    //查询所有数据
    List<AyUser> userList = ayUserService.findAll();
    System.out.println("findAll() :" + userList.size());
    //通过 name 查询数据
    List<AyUser> userList2 = ayUserService.findByName("阿毅");
    System.out.println("findByName() :" + userList2.size());
    Assert.isTrue(userList2.get(0).getName().equals("阿毅"),"data error!");
    //通过 name 模糊查询数据
    List<AyUser> userList3 = ayUserService.findByNameLike("%毅%");
    System.out.println("findByNameLike() :" + userList3.size());
    Assert.isTrue(userList3.get(0).getName().equals("阿毅"),"data error!");
    //通过 ID 列表查询数据
    List<String> ids = new ArrayList<String>();
    ids.add("1");
    ids.add("2");
    List<AyUser> userList4 = ayUserService.findByIdIn(ids);
    System.out.println("findByIdIn() :" + userList4.size());
    //分页查询数据
    PageRequest pageRequest = new PageRequest(0,10);
    Page<AyUser> userList5 = ayUserService.findAll(pageRequest);
    System.out.println("page findAll():" + userList5.getTotalPages() + "/" + userList5.getSize());
    //新增数据
```

```
AyUser ayUser = new AyUser();
ayUser.setId("3");
ayUser.setName("test");
ayUser.setPassword("123");
ayUserService.save(ayUser);
//删除数据
ayUserService.delete("3");
}
```

- Assert：添加Assert断言，在软件开发中是一种常用的调试方式。从理论上来说，通过Assert断言方式可以证明程序的正确性，在现在项目中被广泛使用，这是需要掌握的基本知识。Assert提供了很多好用的方法，比如isNull和isTrue等。

3.3.2 测试

通过运行3.3.1节中的单元测试用例，我们可以在控制台看到如下的打印信息：

```
findAll() :2
findByName() :1
findByNameLike() :1
findByIdIn() :2
page findAll():1/10
```

通过上面的打印信息，可以看出Spring Boot集成Spring Data JPA已经成功，同时代码中的所有Assert断言都全部通过，说明增删改查分页以及自定义查询方法都可以正常运行。

3.4 思考题

什么是Spring Data？

答：Spring Data 的使命是为数据访问提供熟悉且一致的基于 Spring 的编程模型，同时仍保留底层数据存储的特殊特性。它使数据访问技术、关系数据库和非关系数据库、map-reduce 框架和基于云的数据服务变得简单易用。

第 4 章

Thymeleaf 模板引擎与集成测试

本章主要介绍 Thymeleaf 模板引擎、Thymeleaf 模板引擎标签和函数、Spring Boot 中如何使用 Thymeleaf、集成测试以及 Rest Client 工具的使用等内容。

4.1 Thymeleaf 模板引擎介绍

Thymeleaf 是一个优秀的面向 Java 的 XML/XHTML/HTML 5 页面模板，并具有丰富的标签语言和函数。因此，在使用 Spring Boot 框架进行页面设计时，一般都会选择 Thymeleaf 模板。

下面简单列举一下 Thymeleaf 常用的表达式、标签和函数。

常用表达式：

```
${...}     变量表达式
*{...}     选择表达式
#{...}     消息文字表达式
@{...}     链接 url 表达式
#maps      工具对象表达式
```

常用标签：

```
th:action    定义后台控制器路径
th:each      循环语句
th:field     表单字段绑定
th:href      定义超链接
th:id        div 标签中的 id 声明，类似 html 标签中的 id 属性
th:if        条件判断语句
th:include   布局标签，替换内容到引入文件
th:fragment  布局标签，定义一个代码片段，方便其他地方引用
th:object    替换对象
th:src       图片类地址引入
th:text      显示文本
th:value     属性赋值
```

常用函数：

```
#dates       日期函数
#lists       列表函数
#arrays      数组函数
#strings     字符串函数
#numbers     数字函数
#calendars   日历函数
#objects     对象函数
#bools       逻辑函数
```

关于 Thymeleaf 表达式、标签、函数等的更多内容，可以到官方网站 http://www.thymeleaf.org/ 参考学习，这里不过多描述。

4.2 使用 Thymeleaf 模板引擎

4.2.1 引入依赖

要使用 Thymeleaf 模板引擎，需要在 pom.xml 文件中引入如下的依赖（依赖引入之后，记得刷新依赖），具体代码如下：

```xml
<dependency>
    <groupId>org.springframework.boot</groupId>
    <artifactId>spring-boot-starter-thymeleaf</artifactId>
</dependency>
```

同时还需要在 application.properties 文件中添加 Thymeleaf 配置，具体代码如下：

```
#thymeleaf 配置
#模板的模式，支持如 HTML、XML、TEXT、JavaScript 等
spring.thymeleaf.mode=HTML5
#编码，可不用配置
spring.thymeleaf.encoding=UTF-8
#内容类别，可不用配置
spring.thymeleaf.content-type=text/html
#开发配置为 false，避免修改模板还要重启服务器
spring.thymeleaf.cache=false
#配置模板路径，默认就是 templates，可不用配置
#spring.thymeleaf.prefix=classpath:/templates/
```

这里要注意的是，Thymeleaf 模板引擎默认会读取 spring-boot-book-v2 项目下的资源文件夹 resource 下的 templates 目录，这个目录是用来存放 HTML 文件的。如果我们添加了 Thymeleaf 依赖，而没有进行任何配置，或者添加默认目录，启动应用时会报错。

4.2.2 控制层开发

我们在 spring-boot-book-v2 项目目录/src/main/java/com.example.demo.controller 下开发控制层类 AyUserController.java，同时把 AyUserService 服务注入到控制层类当中。具体代码如下：

```java
@Controller
@RequestMapping("/ayUser")
public class AyUserController {

    @Resource
    private AyUserService ayUserService;

    @RequestMapping("/test")
    public String test(Model model) {
        //查询数据库所有用户
        List<AyUser> ayUser = ayUserService.findAll();
```

```
        model.addAttribute("users",ayUser);
        return "ayUser";
    }
}
```

- @Controller：标注此类为一个控制层类，同时让 Spring Boot 容器管理起来。
- @RequestMapping：是一个用来处理请求地址映射的注解，可用于类或者方法上。用于类上，表示类中的所有响应请求的方法都是以该地址作为父路径。@RequestMapping 注解有 value、method 等属性，value 属性可以默认不写。"/ayUser"就是 value 属性的值。value 属性的值就是请求的实际地址。
- Model 对象：一个接口，我们可以把数据库查询出来的数据设置到该类中，前端会从该对象获取数据。其实现类为 ExtendedModelMap，具体可查看源代码：

```
public class ExtendedModelMap extends ModelMap implements Model {
    public ExtendedModelMap() {
    }
}
```

4.2.3 Thymeleaf 模板页面开发

控制层类 AyUserController.java 开发完成之后，我们继续在/src/main/resources/templates 目录下开发 ayUser.html 页面，具体代码如下：

```
<!DOCTYPE HTML>
<html xmlns:th="http://www.thymeleaf.org">
<head>
    <title>hello</title>
    <meta http-equiv="Content-Type" content="text/html; charset=UTF-8" />
</head>
<body>
<table>
    <tr>
        <td>用户名</td>
        <td>密码</td>
    </tr>
    <tr th:each="user:${users}">
        <td th:text="${user.name}"></td>
        <td th:text="${user.password}"></td>
```

```
</tr>
</table>
</body>
</html>
```

<html xmlns:th="http://www.thymeleaf.org">是 Thymeleaf 命名空间，通过引入该命名空间，就可以在 HTML 文件中使用 Thymeleaf 标签语言，用关键字"th"来标注。下面看几个简单的例子：

```
//th:text 用于显示文本 Hello, Thymeleaf
<p th:text=" 'Hello,Thymeleaf' "></p>
//${} 关键字用于获取内存变量为 name 的值
<p th:text="${name}"></p>
//th:src 用于设定<img>图片文件的链接地址   @{} 超链接 url 表达式
<img th:src="@{/image/a.jpg}"/>
```

4.3 集成测试

4.3.1 测试

在 4.2 节中，我们已经简单开发好控制层类和前端 HTML 页面。现在重新运行入口类 MySpringBootApplication 的 main 方法。然后在浏览器中访问 http://localhost:8080/ayUser/test，出现如图 4-1 所示的页面，说明 Spring Boot 集成 Thymeleaf 成功，同时也说明我们开发的前端页面没有问题。

图 4-1　Spring Boot 集成 Thymeleaf 测试

4.3.2　REST Client 工具介绍

REST Client 是一个用于测试 RESTful Web Service 的 Java 客户端。非常小巧，界面也很简单。Intellij IDEA 软件已经集成该插件，方便进行调试。我们可以在 Intellij IDEA 功能菜单中选择【Tools】→【Test RESTful web service】来打开插件。具体如图 4-2 和图 4-3 所示。

图 4-2　进入 REST Client 选项

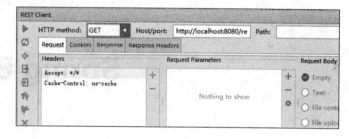
图 4-3　REST Client 展示页面

在 REST Client 插件中，可以在 HTTP method 中选择请求方式：Post 方式或者 Gct 方式等。在 Host/port 中输入需要访问的主机 host 和端口 port，在 Path 中输入访问的映射路径，最后单击右上角的 run 按钮，就可以访问后端代码。同时可以在 Request、Cookies、Response、Response Headers 中查看请求信息、Cookies 信息、响应信息请求头信息等。

4.3.3　使用 REST Client 测试

接下来，我们使用 Rest Clinet 介绍控制层类代码。在 REST Client 界面的 HTTP method 中选择 Get 方式，Host/port 中输入 http://localhost:8080，Path 中输入 ayUser/test，如图 4-4 所示。然后单击 run 按钮运行，当我们看到如图 4-5 所示的 Response 响应界面时，说明代码执行成功。

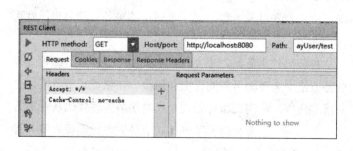
图 4-4　REST Client 测试界面

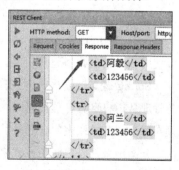
图 4-5　Response 响应页面

第 5 章

Spring Boot 事务支持

本章主要介绍 Spring 声明式事务、Spring 注解事务行为以及在 Spring Boot 中如何使用方法级别事务和类级别事务等。

5.1 Spring 事务介绍

5.1.1 Spring 事务回顾

事务管理是企业级应用程序开发中必不可少的技术，用来确保数据的完整性和一致性。事务有 4 大特性（ACID）：原子性（atomicity）、一致性（consistency）、隔离性（isolation）和持久性（durability）。作为企业级应用程序框架，Spring 在不同的事务管理 API 之上定义了一个抽象层 PlatformTransactionManager，应用程序开发人员不必了解底层的事务管理 API，就可以使用 Spring 的事务管理机制。

Spring 既支持编程式事务管理（也称编码式事务），也支持声明式的事务管理。编程式事务管理是指将事务管理代码嵌入到业务方法中来控制事务的提交和回滚。在编程式事务中，必须在每个业务操作中包含额外的事务管理代码。声明式事务管理是指将事务管理代码从业务方法中分离出来，以声明的方式来实现事务管理。大多数情况下声明式事务管理比编程式事务管理更好用。Spring 通过 Spring AOP 框架支持声明式事务管理。

Spring 并不直接管理事务，而是提供了许多内置事务管理器实现，常用的有 DataSourceTransactionManager、JdoTransactionManager、JpaTransactionManager 以及 HibernateTransactionManager，等等。

5.1.2 Spring 声明式事务

Spring 配置文件中关于事务配置由 3 个组成部分，分别是 DataSource、TransactionManager 和代理机制。无论哪种配置方式，一般变化的只是代理机制部分。DataSource 和 TransactionManager 这两部分只会根据数据访问方式有所变化，比如使用 Hibernate 进行数据访问时，DataSource 实现为 SessionFactory，TransactionManager 的实现为 HibernateTransactionManager。

Spring 声明式事务配置提供 5 种方式，而基于 Annotation 注解方式目前比较流行，所以这里只简单介绍基于注解方式配置 Spring 声明式事务。我们可以使用 @Transactional 注解在类或者方法上表明该类或者方法需要事务支持，被注解的类或者方法被调用时，Spring 开启一个新的事务，当方法正常运行时，Spring 会提交这个事务。具体例子如下：

```
@Transactional
public AyUser updateUser() {
    //执行数据库操作
}
```

这里需要注意的是，@Transactional 注解来自 org.springframework.transaction.annotation。Spring 提供了 @EnableTransactionManagement 注解在配置类上来开启声明式事务的支持。使用 @EnableTransactionManagement 后，Spring 容器会自动扫描注解 @Transactional 的方法和类。

5.1.3 Spring 注解事务行为

当事务方法被另一个事务方法调用时，必须指定事务应该如何传播。例如，方法可能继续在现有事务中运行，也可能开启一个新事务，并在自己的事务中运行。事务的传播行为可以在 @Transactional 的属性中指定。Spring 定义了 7 种传播行为，具体如表 5-1 所示。

表 5-1 Spring 传播行为

传播行为	含义
PROPAGATION_REQUIRED	如果当前没有事务，就新建一个事务，如果已经存在一个事务中，加入到这个事务中
PROPAGATION_SUPPORTS	支持当前事务，如果当前没有事务，就以非事务方式执行

(续表)

传播行为	含 义
PROPAGATION_MANDATORY	使用当前的事务,如果当前没有事务,就抛出异常
PROPAGATION_REQUIRES_NEW	新建事务,如果当前存在事务,把当前事务挂起
PROPAGATION_NOT_SUPPORTED	以非事务方式执行操作,如果当前存在事务,就把当前事务挂起
PROPAGATION_NEVER	以非事务方式执行,如果当前存在事务,则抛出异常
PROPAGATION_NESTED	如果当前存在事务,则在嵌套事务内执行。如果当前没有事务,则执行与 PROPAGATION_REQUIRED 类似的操作

隔离级别定义了一个事务可能受其他并发事务影响的程度。在典型的应用程序中,多个事务并发运行,经常会操作相同的数据来完成各自的任务。并发虽然是必需的,但也可是会导致许多问题,并发事务所导致的问题可以分为以下 3 类。

- 脏读(Dirty reads):脏读发生在一个事务读取了另一个事务改写但尚未提交的数据。如果改写在稍后被回滚了,那么第一个事务获取的数据就是无效的。
- 不可重复读(Nonrepeatable read):不可重复读发生在一个事务执行相同的查询两次或两次以上,但是每次都得到不同的数据时。这通常是因为另一个并发事务在两次查询期间更新了数据。
- 幻读(Phantom read):幻读与不可重复读类似。它发生在一个事务(T1)读取了几行数据,接着另一个并发事务(T2)插入了一些数据时。在随后的查询中,第一个事务(T1)就会发现多了一些原本不存在的记录。

针对这些问题,Spring 提供了 5 种事务的隔离级别,具体如表 5-2 所示。

表 5-2 Spring 隔离级别

隔离级别	含 义
ISOLATION_DEFAULT	使用数据库默认的事务隔离级别,另外 4 个与 JDBC 的隔离级别相对应
ISOLATION_READ_UNCOMMITTED	事务最低的隔离级别,允许读取尚未提交的更改。可能导致脏读、幻读或不可重复读
ISOLATION_READ_COMMITTED	允许从已经提交的并发事务读取。可防止脏读,但幻读和不可重复读仍可能会发生
ISOLATION_REPEATABLE_READ	对相同字段的多次读取的结果是一致的,除非数据被当前事务本身改变。可防止脏读和不可重复读,但幻读仍可能发生
ISOLATION_SERIALIZABLE	完全服从 ACID 的隔离级别,确保不发生脏读、不可重复读和幻读。这种隔离级别是最慢的,因为它通常是通过完全锁定当前事务所涉及的数据表来完成的

@Transactional 可以通过 propagation 属性定义事务行为，属性值分别为 REQUIRED、SUPPORTS、MANDATORY、REQUIRES_NEW、NOT_SUPPORTED、NEVER 以及 NESTED，分别对应表 5-1 中的内容。可以通过 isolation 属性定义隔离级别，属性值分别为 DEFAULT、READ_UNCOMMITTED、READ_COMMITTED、REPEATABLE_READ 以及 SERIALIZABLE。

还可以通过 timeout 属性设置事务过期时间，通过 readOnly 指定当前事务是否是只读事务，通过 rollbackFor（noRollbackFor）指定哪个或者哪些异常可以引起（或不可以引起）事务回滚。

5.2 Spring Boot 事务使用

5.2.1 Spring Boot 事务介绍

Spring Boot 开启事务很简单，只需要一个注解@Transactional 就可以了，因为在 Spring Boot 中已经默认对 JPA、JDBC、MyBatis 开启了事务，引入它们依赖的时候，事物就默认开启。当然，如果你需要用其他的 ORM 框架，比如 BeatlSQL，就需要自己配置相关的事务管理器。

Spring Boot 用于配置事务的类为 TransactionAutoConfiguration，此配置类依赖于 JtaAutoConfiguration 和 DataSourceTransactionManagerAutoConfiguration，具体查看源代码可知，而 DataSourceTransactionManagerAutoConfiguration 已开启了对声明式事务的支持，所以在 Spring Boot 中，无须显示开启使用@EnableTransactionManagement。

5.2.2 类级别事务

在第 2 章中，我们已经在 Spring Boot 中集成了 Spring Data JPA，同时开发了 AyUserRepository 类实现 JpaRepository 接口，JpaRepository 接口是不开启事务的，而 SimpleJapRepository 默认是开启事务的，所以我们需要手工给 AyUserRepository 添加事务。AyUserRepository 类中的方法是在服务层类 AyUserServiceImpl 中被使用，而事务一般都是加在服务层，因此可以在 AyUserServiceImpl 类上添加@Transactional 注解来开启事务。AyUserServiceImpl 类开启事务的代码如下：

```
/**
 * 描述：用户服务层实现类
 * @author 阿毅
```

```
 * @date    2017/10/14
 */
@Transactional
@Service
public class AyUserServiceImpl implements AyUserService {

    @Resource(name = "ayUserRepository")
    private AyUserRepository ayUserRepository;

    //省略代码
}
```

@Transactional 注解在类上，意味着此类的所有 public 方法都是开启事务的。

5.2.3 方法级别事务

@Transactional 除了可以注解在类上，还可以注解到方法上面。当注解在类上的时候意味着此类的所有 public 方法都是开事务的。如果类级别和方法级别同时使用了@Transactional 注解，则使用方法级别注解覆盖类级别注解。可以给 AyUserServiceImpl 类中的 save()方法添加事务，同时在 save 完成之后抛出 NullPointException 异常，查看数据是否可以回滚，具体代码如下：

```
/**
 * 描述：用户服务层实现类
 * @author 阿毅
 * @date    2017/10/14
 */
//注解在类上
@Transactional
@Service
public class AyUserServiceImpl implements AyUserService {

    @Resource(name = "ayUserRepository")
    private AyUserRepository ayUserRepository;

    //注解在方法上
    @Transactional
    @Override
```

```
public AyUser save(AyUser ayUser) {
    AyUser saveUser = ayUserRepository.save(ayUser);
    //出现空指针异常
    String error = null;
    error.split("/");
    return saveUser;
}


}
```

5.2.4 测试

5.2.1 节和 5.2.2 节代码开发完成之后，我们在测试类 DemoApplicationTests 中添加测试方法，具体代码如下：

```
@Test
public void testTransaction(){
    AyUser ayUser = new AyUser();
    ayUser.setId("3");
    ayUser.setName("阿华");
    ayUser.setPassword("123");
    ayUserService.save(ayUser);
}
```

运行 testTransaction()单元测试用例，当代码执行完成后，由于方法 save 保持数据时，出现空指针，数据会回滚，数据库查询不到保存的数据。现在我们把 AyUserServiceImpl 类上的 @Transactional 注解和 save 方法上的@ Transactional 注解全部注释掉，再次执行 testTransaction() 单元测试用例，查询数据库，发现数据库多了一条数据，如图 5-1 所示。

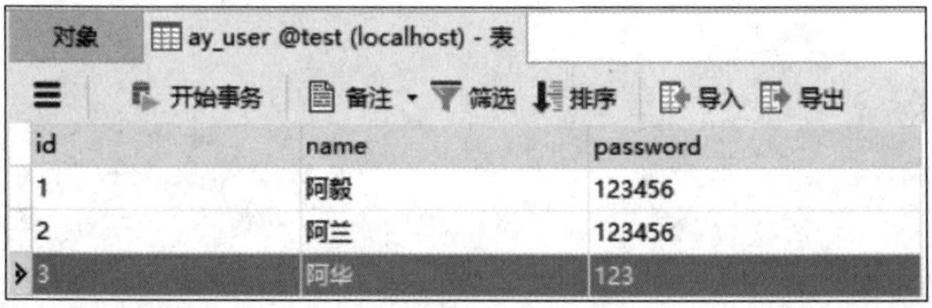

id	name	password
1	阿毅	123456
2	阿兰	123456
3	阿华	123

图 5-1 数据插入到数据库

5.3 思考题

微服务同时调用多个接口时,是怎么支持事务的?

答:可以使用 Spring Boot 集成 Aatomikos 来解决分布式事务,但是笔者一般不建议这样使用,因为使用分布式事务会增加请求的响应时间,影响系统的 TPS。一般在实际工作中,会利用消息的补偿机制来处理分布式的事务。

第 6 章

使用过滤器和监听器

本章主要介绍如何在 Spring Boot 中使用过滤器 Filter 和监听器 Listener。

6.1　Spring Boot 使用过滤器 Filter

6.1.1　过滤器 Filter 介绍

过滤器英文名称为 Filter，是处于客户端与服务器资源文件之间的一道过滤网，它是 Servlet 技术中最激动人心的技术之一。Web 开发人员通过 Filter 技术管理 Web 服务器的所有资源，例如 JSP、Servlet、静态图片文件或静态 HTML 文件等进行拦截，从而实现一些特殊的功能。例如实现 URL 级别的权限访问控制、过滤敏感词汇、压缩响应信息等一些高级功能。

Filter 接口源代码如下：

```
public interface Filter {
    void init(FilterConfig var1) throws ServletException;
    void doFilter(ServletRequest var1, ServletResponse var2, FilterChain var3) throws
IOException, ServletException;
    void destroy();
}
```

Filter 的创建和销毁由 Web 服务器负责。Web 应用程序启动时，Web 服务器将创建 Filter 的实例对象，并调用其 init 方法，读取 web.xml 配置，完成对象的初始化功能，从而为后续的用户请求作好拦截的准备工作（Filter 对象只会创建一次，init 方法也只会执行一次）。开发人员通过 init 方法的参数，可获得代表当前 filter 配置信息的 FilterConfig 对象。

当客户请求访问与过滤器关联的 URL 时，过滤器将先执行 doFilter 方法。FilterChain 参数用于访问后续过滤器。Filter 对象创建后会驻留在内存，当 Web 应用移除或服务器停止时才销毁。在 Web 容器卸载 Filter 对象之前 destroy 被调用。该方法在 Filter 的生命周期中仅执行一次。在这个方法中，可以释放过滤器使用的资源。

Filter 可以是有很多个，当一个个 Filter 组合成起来，就形成了一个 FilterChain，也就是我们说的过滤链，具体如图 6-1 所示。

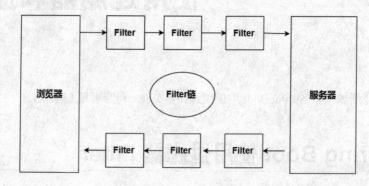

图 6-1　Filter 链

FilterChain 的执行顺序遵循先进后出的原则：当 Web 客户端发送一个 Request 请求的时候，这个 Request 请求会先经过 FilterChain，由它利用 dofilter()方法调用各个子 Filter，至于子 filter 的执行顺序如何，则要看客户端是如何制定规则的。当 Request 请求被第一个 Filter 处理之后，又通过 dofilte()往下传送，被第二个、第三个……Filter 截获处理。当 Request 请求被所有的 Filter 处理之后，返回的顺序是从最后一个开始进行返回，直到返回给客户端。

6.1.2　过滤器 Filter 的使用

在 Spring Boot 中使用 Filter 过滤器很简单。首先在项目 spring-boot-book-v2 的目录 /src/main/java/com.example.demo.filter 下新建 AyUserFilter.java 类，具体代码如下：

```
/**
 * 过滤器
 * @author Ay
```

```java
 * @date    2017/11/2
 */
@WebFilter(filterName = "ayUserFilter", urlPatterns = "/*")
public class AyUserFilter implements Filter{

    @Override
    public void init(FilterConfig filterConfig) throws ServletException {
        System.out.println("---------->>> init");
    }

    @Override
    public void doFilter(ServletRequest servletRequest, ServletResponse servletResponse,
        FilterChain filterChain) throws IOException, ServletException {
        System.out.println("---------->>> doFilter");
        filterChain.doFilter(servletRequest,servletResponse);
    }

    @Override
    public void destroy() {
        System.out.println("---------->>> destory");
    }
}
```

- @WebFilter: 用于将一个类声明为过滤器，该注解将会在应用部署时被容器处理，容器根据具体的属性配置将相应的类部署为过滤器。这样在Web应用中使用监听器时，不需要在web.xml文件中配置监听器的相关描述信息了。该注解的常用属性有: filterName、urlPatterns、value等。filterName属性用于指定过滤器的name，等价于XML配置文件中的<filter-name>标签。urlPatterns属性用于指定一组过滤器的URL匹配模式，等价于XML配置文件中的<url-pattern>标签。value属性等价于urlPatterns属性，但是两者不可以同时使用。

AyUserFilter.java 类开发完成之后，我们需要在入口类 DemoApplication.java 中添加注解 @ServletComponentScan，具体代码如下:

```java
@SpringBootApplication
@ServletComponentScan
public class MySpringBootApplication {

    public static void main(String[] args) {
```

```
        SpringApplication.run(MySpringBootApplication.class, args);
    }
}
```

- @ServletComponentScan：使用该注解后，Servlet、Filter、Listener可以直接通过@WebServlet、@WebFilter、@WebListener注解自动注册，无须其他代码。

事实上，在 Spring Boot 中添加自己的 Servlet、Filter 和 Listener 有两种方法，代码注册和注解自动注册。上面已经讲解了如何用注解自动注册，而代码注册可以通过 ServletRegistrationBean、FilterRegistrationBean 和 ServletListenerRegistrationBean 注册 Bean。虽然条条大路通罗马，但是希望大家先掌握一种方式，一路走到底，而不是纠结于会有几种写法。

6.1.3 测试

6.1.2 节代码开发完成之后，重新启动运行 spring-boot-book-v2 项目时，Web 容器会初始化 AyUserFilter 对象，并调用 init 方法，可以在 Intellij IDEA 控制台看到打印信息，如图 6-2 所示。在浏览器输入 http://localhost:8080/ayUser/test 访问应用时，AyUserFilter 拦截器会拦截本次的请求，并调用 doFilter 方法，同时会在控制台打印信息，如图 6-3 所示。

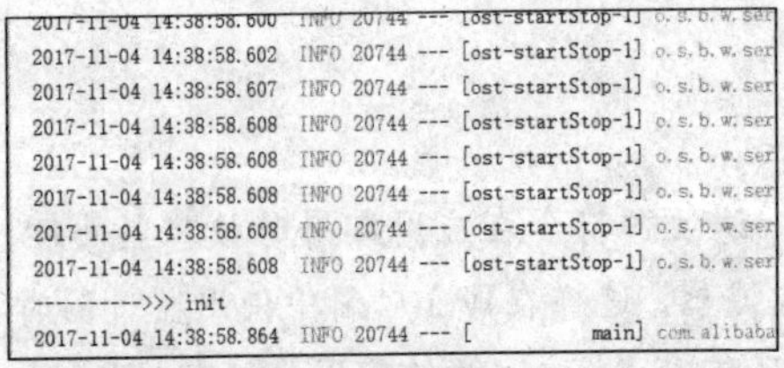

图 6-2　初始化调用 init 方法打印信息 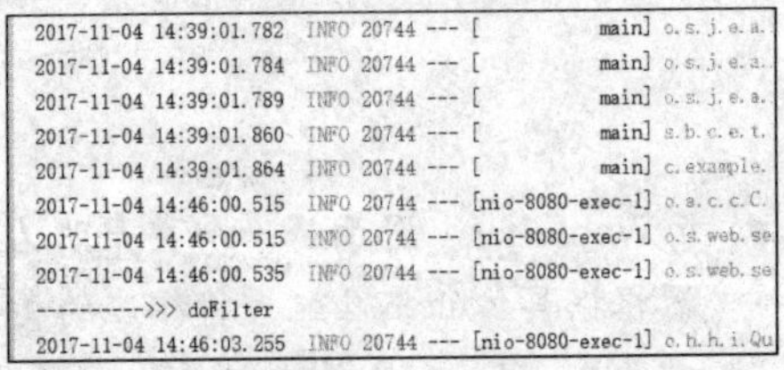　　　　图 6-3　拦截调用 doFilter 方法打印信息

6.2　Spring Boot 使用监听器 Listener

6.2.1　监听器 Listener 介绍

监听器也叫 Listener，是 Servlet 的监听器，它可以用于监听 Web 应用中某些对象、信息的创建、销毁、增加、修改、删除等动作的发生，然后做出相应的响应处理。当范围对象的状态发生变化的时候，服务器自动调用监听器对象中的方法。常用于统计在线人数和在线用户，系统加载时进行信息初始化，统计网站的访问量，等等。

根据监听对象，可把监听器分为 3 类：ServletContext（对应 application）、HttpSession（对应 session）、ServletRequest（对应 request）。Application 在整个 Web 服务中只有一个，在 Web 服务关闭时销毁。Session 对应每个会话，在会话起始时创建，一端关闭会话时销毁。Request 对象是客户发送请求时创建的（一同创建的还有 Response），用于封装请求数据，在一次请求处理完毕时销毁。

根据监听事件分为监听对象创建与销毁，如 ServletContextListener、监听对象域中属性的增加和删除，如：HttpSessionListener 和 ServletRequestListener、监听绑定到 Session 上的某个对象的状态，如 ServletContextAttributeListener、HttpSessionAttributeListener 和 ServletRequestAttributeListener 等。

6.2.2 监听器 Listener 的使用

在 Spring Boot 中使用 Listener 监听器和 Filter 基本一样。首先在项目 spring-boot-book-v2 的目录/src/main/java/com.example.demo.listener 下新建 AyUserListener.java 类，具体代码如下：

```java
/**
 * 描述：监听器
 * @author Ay
 * @date    2017/11/4
 */
@WebListener
public class AyUserListener implements ServletContextListener {

    @Override
    public void contextInitialized(ServletContextEvent servletContextEvent) {
        System.out.println("ServletContext 上下文初始化");
    }

    @Override
    public void contextDestroyed(ServletContextEvent servletContextEvent) {
        System.out.println("ServletContext 上下文销毁");
    }

}
```

- @WebListener：用于将一个类声明为监听器，该注解将会在应用部署时被容器处理，容器根据具体的属性配置将相应的类部署为监听器。这样在Web应用中使用监听器时，不需要在web.xml文件中配置监听器的相关描述信息。

- ServletContextListener类：能够监听ServletContext对象的生命周期，实际上就是监听Web应用的生命周期。当Servlet 容器启动或终止Web应用时，会触发ServletContextEvent 事件，该事件由ServletContextListener类来处理。在ServletContextListener接口中定义了处理ServletContextEvent 事件的两个方法：contextInitialized和contextDestroyed。
  - contextInitialized：当Servlet容器启动Web 应用时调用该方法。在调用完该方法之后，容器再对Filter 初始化，并且对那些在Web 应用启动时就需要被初始化的Servlet进行初始化。
  - contextDestroyed：当Servlet容器终止Web应用时调用该方法。在调用该方法之前，容器会销毁所有的Servlet和Filter过滤器。

我们可以在 contextInitialized 方法中查询所有的用户，利用缓存技术把用户数据存放到缓存中。在第 7 章中我们会具体讲解如何利用监听器和 Redis 缓存技术来缓存用户数据，提高系统性能。

6.2.3 测试

6.2.2 节代码开发完成之后，重新启动运行 spring-boot-book-v2 项目时，Web 容器会初始化 AyUserListener 对象并调用 contextInitialized 方法，可以在 Intellij IDEA 控制台看到打印信息，如图 6-4 所示。当我们销毁容器时，会调用 contextDestroyed 方法并在控制台打印信息。这里需要注意的是，在 IDEA 开发工具中，直接终止容器或者干掉进程是不会执行销毁方法 contextDestroyed 的。

```
2017-11-04 18:40:41.875  INFO 23484 --- [ost-startStop-1] o.s.b.w.servlet.S
2017-11-04 18:40:41.879  INFO 23484 --- [ost-startStop-1] o.s.b.w.servlet.F
2017-11-04 18:40:41.880  INFO 23484 --- [ost-startStop-1] o.s.b.w.servlet.F
2017-11-04 18:40:41.880  INFO 23484 --- [ost-startStop-1] o.s.b.w.servlet.F
2017-11-04 18:40:41.880  INFO 23484 --- [ost-startStop-1] o.s.b.w.servlet.F
2017-11-04 18:40:41.880  INFO 23484 --- [ost-startStop-1] o.s.b.w.servlet.F
2017-11-04 18:40:41.880  INFO 23484 --- [ost-startStop-1] o.s.b.w.servlet.F
ServletContext上下文初始化
```

图 6-4 容器启动监听打印信息

第 7 章

集成 Redis 缓存

本章主要介绍如何安装 Redis 缓存、Redis 缓存 5 种基本数据类型的增删改查、Spring Boot 中如何集成 Redis 缓存以及如何使用 Redis 缓存用户数据等内容。

7.1 Redis 缓存介绍

7.1.1 Redis 概述

Redis 是一个基于内存的单线程高性能 key-value 型数据库，读写性能优异。和 Memcached 缓存相比，Redis 支持丰富的数据类型，包括 string（字符串）、list（链表）、set（集合）、zset（sorted set 有序集合）和 hash（哈希类型）。因此 Redis 在企业中被广泛使用。

7.1.2 Redis 服务器安装

Redis 项目本身是不支持 Windows 的，但是 Microsoft 开发技术小组针对 Win64 开发和维护了 Windows 端口，所以可以在网络上下载 Redis 的 Windows 版本。具体步骤如下：

步骤 01 打开官方网站 http://redis.io/，单击 Download，如图 7-1 所示。

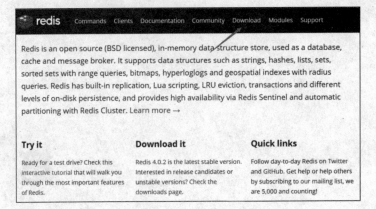

图 7-1 Redis 下载首页

步骤 02 在弹出的页面中，找到 Learn more 选项，并单击进入，具体如图 7-2 所示。

步骤 03 在弹出的页面中选择【releases】选项，具体如图 7-3 所示。

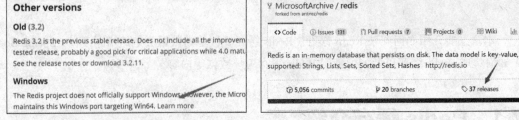

图 7-2 单击 Learn more 链接　　　　图 7-3 选择 Download ZIP 下载 Redis

步骤 04 在弹出的界面中选择 Redis 3.0.504 这个版本，选择其他版本也可以，单击【Redis-x64-3.0.504.zip】下载 Redis 安装包，如图 7-4 所示。

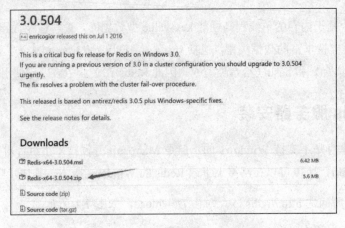

图 7-4 下载 Redis 3.0.504 安装包

步骤 05 解压下载的安装包【Redis-x64-3.0.504.zip】，双击【redis-server.exe】，Redis 服务器就运行起来了，如图 7-5 所示。同时可以看到 Redis 启动成功的界面，如图 7-6 所示。

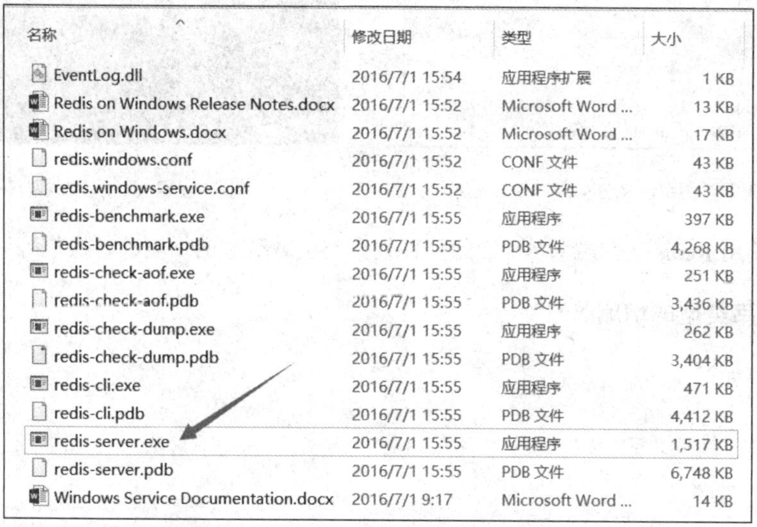

图 7-5 启动 Redis 服务器

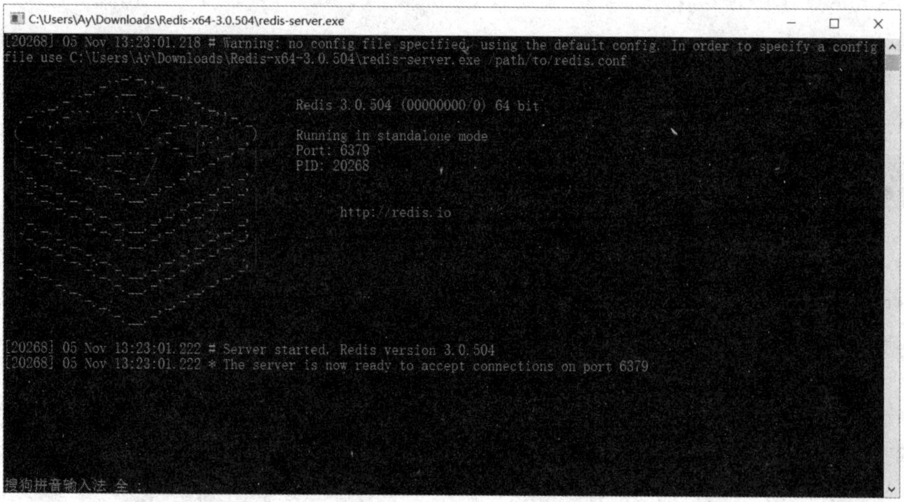

图 7-6 Redis 启动成功界面

7.1.3 Redis 缓存测试

Redis 安装成功之后，可以在安装包里找到 Redis 客户端程序 redis-cli.exe，如图 7-7 所示，双击 redis-cli.exe，打开 Redis 客户端界面，如图 7-8 所示。

图 7-7 启动 Redis 客户端　　图 7-8 Redis 启动成功界面

下面就使用 Redis 客户端对 Redis 的几种数据类型做基本的增删改查操作练习。

1. 字符串类型的增删改查

具体代码如下：

```
###增加一个值 key 为 name，value 为 ay
127.0.0.1:6379> set name 'ay'
OK
###查询 name 的值
127.0.0.1:6379> get name
"ay"
###更新 name 的值为 al
127.0.0.1:6379> set name 'al'
OK
###查询 name 的值
127.0.0.1:6379> get name
"al"
###删除 name 的值
127.0.0.1:6379> del name
(integer) 1
###查询是否存在 name，0 代表不存在
127.0.0.1:6379> exists name
(integer) 0
127.0.0.1:6379>
```

2. List 集合的增删改查

具体代码如下：

```
###添加 key 为 user_list，value 为 'ay'，'al' 的 list 集合
127.0.0.1:6379> lpush user_list 'ay' 'al'
```

```
(integer) 2
###查询key为user_list的集合
127.0.0.1:6379> lrange user_list 0 -1
1) "al"
2) "ay"
###往list尾部添加love元素
127.0.0.1:6379> rpush user_list 'love'
(integer) 3
###往list头部添加hope元素
127.0.0.1:6379> lpush user_list 'hope'
(integer) 4
###查询key为user_list的集合
127.0.0.1:6379> lrange user_list 0 -1
1) "hope"
2) "al"
3) "ay"
4) "love"
###更新index为0的值
127.0.0.1:6379> lset user_list 0 'wish'
OK
###查询key为user_list的集合
127.0.0.1:6379> lrange user_list 0 -1
1) "wish"
2) "al"
3) "ay"
4) "love"
###删除index为0的值
127.0.0.1:6379> lrem user_list 0 'wish'
(integer) 1
###查询key为user_list的集合
127.0.0.1:6379> lrange user_list 0 -1
1) "al"
2) "ay"
3) "love"
127.0.0.1:6379>
```

3. Set集合的增删改查

具体代码如下：

```
###添加key为user_set,value为"ay" "al" "love"的集合
127.0.0.1:6379> sadd user_set "ay" "al" "love"
(integer) 3
###查询key为user_set集合
127.0.0.1:6379> smembers user_set
1) "al"
2) "ay"
3) "love"
###删除value为love，返回1表示删除成功，0表示失败
127.0.0.1:6379> srem user_set 'love'
(integer) 1
###查询set集合所有值
127.0.0.1:6379> smembers user_set
1) "al"
2) "ay"
###添加love元素，set集合是没有顺序的，所以无法判断添加到哪个位置
127.0.0.1:6379> sadd user_set 'love'
(integer) 1
###查询set集合所有值，发现添加到第二个位置
127.0.0.1:6379> smembers user_set
1) "al"
2) "love"
3) "ay"
###添加love元素，由于set集合已经存在，返回0代表添加不成功，但是不会报错
127.0.0.1:6379> sadd user_set 'love'
(integer) 0
```

4. Hash集合的增删改查

具体代码如下：

```
###清除数据库
127.0.0.1:6379> flushdb
OK
###创建hash，key为user_hset,字段为user1，值为ay
127.0.0.1:6379> hset user_hset "user1" "ay"
```

```
(integer) 1
```
往 key 为 user_hset 添加字段为 user2，值为 al
```
127.0.0.1:6379> hset user_hset "user2" "al"
(integer) 1
```
查询 user_hset 字段长度
```
127.0.0.1:6379> hlen user_hset
(integer) 2
```
查询 user_hset 所有字段
```
127.0.0.1:6379> hkeys user_hset
1) "user1"
2) "user2"
```
查询 user_hset 所有值
```
127.0.0.1:6379> hvals user_hset
1) "ay"
2) "al"
```
查询字段 user1 的值
```
127.0.0.1:6379> hget user_hset "user1"
"ay"
```
获取 key 为 user_hset 所有的字段和值
```
127.0.0.1:6379> hgetall user_hset
1) "user1"
2) "ay"
3) "user2"
4) "al"
```
更新字段 user1 的值为 new_ay
```
127.0.0.1:6379> hset user_hset "user1" "new_ay"
(integer) 0
```
更新字段 user2 的值为 new_al
```
127.0.0.1:6379> hset user_hset "user2" "new_al"
(integer) 0
```
获取 key 为 user_hset 所有的字段和值
```
127.0.0.1:6379> hgetall user_hset
1) "user1"
2) "new_ay"
3) "user2"
4) "new_al"
```
删除字段 user1 和值
```
127.0.0.1:6379> hdel user_hset user1
```

```
(integer) 1
###获取 key 为 user_hset 所有的字段和值
127.0.0.1:6379> hgetall user_hset
1) "user2"
2) "new_al"
127.0.0.1:6379>
```

5. SortedSet集合的增删改查

具体代码如下：

```
###清除数据库
127.0.0.1:6379> flushdb
OK
###SortedSet 集合添加 ay 元素，分数为 1
127.0.0.1:6379> zadd user_zset 1 "ay"
(integer) 1
###SortedSet 集合添加 al 元素，分数为 2
127.0.0.1:6379> zadd user_zset 2 "al"
(integer) 1
###SortedSet 集合添加 love 元素，分数为 3
127.0.0.1:6379> zadd user_zset 3 "love"
(integer) 1
###按照分数由小到大查询 user_zset 集合元素
127.0.0.1:6379> zrange user_zset 0 -1
1) "ay"
2) "al"
3) "love"
###按照分数由大到小查询 user_zset 集合元素
127.0.0.1:6379> zrevrange user_zset 0 -1
1) "love"
2) "al"
3) "ay"
###查询元素 ay 的分数值
127.0.0.1:6379> zscore user_zset "ay"
"1"
###查询元素 love 的分数值
127.0.0.1:6379> zscore user_zset "love"
"3"
```

7.2　Spring Boot 集成 Redis 缓存

7.2.1　Spring Boot 缓存支持

在 Spring Boot 中提供了强大的基于注解的缓存支持，可以通过注解配置的方式低侵入地给原有 Spring 应用增加缓存功能，提高数据访问的性能。Spring Boot 为我们配置了多个 CacheManager 的实现，可以根据具体的项目要求使用相应的缓存技术，如图 7-9 所示。

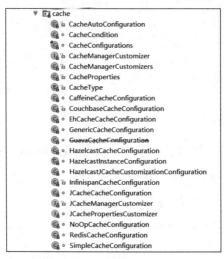

图 7-9　Spring Boot 缓存配置类

从图 7-9 可知，Spring Boot 支持许多类型的缓存，比如 EhCache、JCache、Redis 等。在不添加任何额外配置的情况下，Spring Boot 默认使用 SimpleCacheConfiguration，考虑到 Redis 缓存在企业中被广泛使用，故选择用 Redis 缓存来进行讲解。

7.2.2　引入依赖

在 Spring Boot 中集成 Redis，首先需要在 pom.xml 文件中引入所需的依赖，具体代码如下：

```
<dependency>
        <groupId>org.springframework.boot</groupId>
        <artifactId>spring-boot-starter-data-redis</artifactId>
</dependency>
```

7.2.3 添加缓存配置

在 pom 文件引入 Redis 所需的依赖之后,需要在 application.properties 文件中添加如下的配置信息:

```
### redis 缓存配置
### 默认 redis 数据库为 db0
spring.redis.database=0
### 服务器地址,默认为 localhost
spring.redis.host=localhost
### 链接端口,默认为 6379
spring.redis.port=6379
### redis 密码默认为空
spring.redis.password=
```

7.2.4 测试用例开发

在 application.properties 配置文件中添加完 Redis 配置之后,在测试类 DemoApplicationTests.java 中继续添加如下的代码:

```
@Resource
    private RedisTemplate redisTemplate;

    @Resource
    private StringRedisTemplate stringRedisTemplate;

    @Test
    public void testRedis(){
        //增 key: name, value: ay
        redisTemplate.opsForValue().set("name","ay");
        String name = (String)redisTemplate.opsForValue().get("name");
        System.out.println(name);
        //删除
        redisTemplate.delete("name");
        //更新
        redisTemplate.opsForValue().set("name","al");
        //查询
        name = stringRedisTemplate.opsForValue().get("name");
```

```
System.out.println(name);
}
```

RedisTemplate 和 StringRedisTemplate 都是 Spring Data Redis 提供的两个模板类用来对数据进行操作，其中 StringRedisTemplate 只针对键值都是字符串的数据进行操作。在应用启动的时候，Spring 会初始化这两个模板类，通过@Resource 注解注入即可使用。

RedisTemplate 和 StringRedisTemplate 除了提供 opsForValue 方法用来操作简单属性数据之外，还提供了以下数据访问方法。

（1）opsForList：操作含有 list 的数据。
（2）opsForSet：操作含有 set 的数据。
（3）opsForZSet：操作含有 ZSet（有序 set）的数据。
（4）opsForHash：操作含有 hash 的数据。

当我们的数据存放到 Redis 的时候，键（key）和值（value）都是通过 Spring 提供的 Serializer 序列化到数据库的。RedisTemplate 默认使用 JdkSerializationRedisSerializer，而 StringRedisTemplate 默认使用 StringRedisSerializer。

7.2.5 测试

7.2.4 节测试用例代码开发完成之后，运行单元测试用例，除了可以在控制台查看打印结果信息和在 Redis 客户端查看数据之外，还可以使用 RedisClient 客户端工具查看 Redis 缓存数据库中的数据。RedisClient 客户端软件可以到网络上下载，并安装到自己的操作系统中。安装完成之后，可以看到如图 7-10 所示的界面。

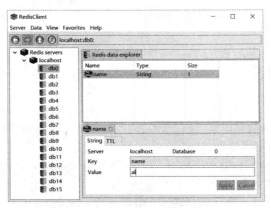

图 7-10　RedisClient 界面

在图 7-10 中，可以看到 Redis 默认有 16 个数据库，客户端与 Redis 建立连接后会自动选择 0 号数据库。通过该客户端可以清楚地查看 Redis 数据库中存放的数据情况，同时可以在客户端中对数据进行增删改查等操作，方便使用。

7.3 Redis 缓存在 Spring Boot 中的使用

7.3.1 监听器 Listener 开发

在 6.2.2 节当中，我们已经简单地开发好 AyUserListener 监听器类，并在上下文启动时打印信息。在本节中，我们想在上下文初始化的方法中，加载数据库中的所有用户数据，并存放到 Redis 缓存中。之所以要把用户数据存放到缓存中，是因为用户的数据属于变动不大的数据，适合存放到缓存中，在应用需要获取用户数据时，可以直接到 Redis 缓存中获取，不用到数据库中获取数据库连接查询数据，提高数据的访问速度。具体代码如下：

```
/**
 * 描述：监听器
 * @author Ay
 * @date   2017/11/4
 */
@WebListener
public class AyUserListener implements ServletContextListener {

    @Resource
    private RedisTemplate redisTemplate;
    @Resource
    private AyUserService ayUserService;
    private static final String ALL_USER = "ALL_USER_LIST";

    @Override
    public void contextInitialized(ServletContextEvent servletContextEvent) {
        //查询数据库所有的用户
        List<AyUser> ayUserList = ayUserService.findAll();
        //清除缓存中的用户数据
        redisTemplate.delete(ALL_USER);
        //将数据存放到 Redis 缓存中
```

```
        redisTemplate.opsForList().leftPushAll(ALL_USER, ayUserList);
        //真实项目中需要注释掉,查询所有的用户数据
        List<AyUser> queryUserList = redisTemplate.opsForList().range(ALL_USER, 0, -1);
        System.out.println("缓存中目前的用户数有: " + queryUserList.size() + " 人");
        System.out.println("ServletContext 上下文初始化");
    }

    @Override
    public void contextDestroyed(ServletContextEvent servletContextEvent) {
        System.out.println("ServletContext 上下文销毁");
    }
}
```

- redisTemplate.opsForList().leftPushAll:查询缓存中所有的用户数据,ALL_USER键若不存在,会创建该键及与其关联的List,之后再将参数中的ayUserList从左到右依次插入。
- rcdisTemplate.opsForList().range:取链表中的全部元素,其中0表示第一个元素,-1表示最后一个元素。

在 7.2.3 节中已经提到,当我们的数据存放到 Redis 的时候,键(key)和值(value)都是通过 Spring 提供的 Serializer 序列化到数据库的。RedisTemplate 默认使用 JdkSerializationRedisSerializer,而 StringRedisTemplate 默认使用 StringRedisSerializer。所以我们需要让用户类 AyUser(/src/main/java/com.example.demo.model)实现序列化接口 Serializable,具体代码如下:

```
/**
 * 描述: 用户表
 * @Author 阿毅
 * @date   2017/10/8
 */
@Entity
@Table(name = "ay_user")
public class AyUser implements Serializable{
    //省略代码
}
```

7.3.2 项目启动缓存数据

在 7.3.1 节中，已经开发好 AyUserListener 监听器类和 AyUser 用户类，重新启动项目，这时候数据库表 ay_user 中的所有数据都会加载到 Redis 缓存中。在 contextInitialized 方法中进行断点调试，出现如图 7-11 所示的图片，可见代码数据已经成功地被加载到缓存中。我们也可以用 RedisClient 客户端软件来查看用户数据是否存放到缓存中。

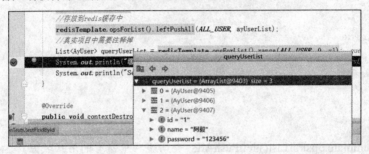

图 7-11　Redis 断点调试界面

7.3.3 更新缓存数据

项目启动加载所有用户数据到缓存之后，需要修改 AyUserServiceImpl 中的接口，比如 findById、Save、Delete 等方法。因为如果在 Redis 缓存中查询不到数据，那么就需要到数据库中查询，如果能够在数据库中查询到数据，除了返回数据之外，还需要把数据更新到缓存中。这样再次查询数据时，就不需要到数据库中查询数据。这里主要对方法 findById 进行修改，AyUserServiceImpl 中具体需要修改的代码如下：

```
//省略代码
@Service
public class AyUserServiceImpl implements AyUserService {

    @Resource(name = "ayUserRepository")
    private AyUserRepository ayUserRepository;

    @Resource
    private RedisTemplate redisTemplate;

    private static final String ALL_USER = "ALL_USER_LIST";
    @Override
```

```java
public AyUser findById(String id){
    //step.1 查询Redis缓存中的所有数据
    List<AyUser> ayUserList = redisTemplate.opsForList().range(ALL_USER, 0, -1);
    if(ayUserList != null && ayUserList.size() > 0){
        for(AyUser user : ayUserList){
            if (user.getId().equals(id)){
                return user;
            }
        }
    }
    //step.2 查询数据库中的数据
    AyUser ayUser = ayUserRepository.findById(id).get();
    if(ayUser != null){
        //step.3 将数据插入到Redis缓存中
        redisTemplate.opsForList().leftPush(ALL_USER, ayUser);
    }
    return ayUser;
}
//省略代码
}
```

对于 Save、Delete 等方法的修改，思路是一样的，这里就不一一重复赘述，读者可自己尝试实现。虽然引入 Redis 缓存用户数据可以提高访问性能，但是带来的代码复杂度也是可想而知。所以在工作中，需要权衡性能和代码的复杂度，根据具体的业务场景加以选择，不可滥用缓存。

7.3.4 测试

7.3.3 节代码开发完成之后，在测试类 MySpringBootApplicationTests 下继续添加如下的测试方法：

```java
@Test
public void testFindById(){
    Long redisUserSize = 0L;
    //查询id = 1 的数据，该数据存在于Redis缓存中
    AyUser ayUser = ayUserService.findById("1");
```

```java
            redisUserSize = redisTemplate.opsForList().size("ALL_USER_LIST");
            System.out.println("目前缓存中的用户数量为: " + redisUserSize);
            System.out.println("--->>> id: " + ayUser.getId() + " name:" + ayUser.getName());
            //查询id = 2 的数据，该数据存在于Redis 缓存中
            AyUser ayUser1 = ayUserService.findById("2");
            redisUserSize = redisTemplate.opsForList().size("ALL_USER_LIST");
            System.out.println("目前缓存中的用户数量为: " + redisUserSize);
            System.out.println("--->>> id: " + ayUser1.getId() + " name:" + ayUser1.getName());
            //查询id = 4 的数据，不存在于Redis 缓存中，存在于数据库中，所以会把数据库查询的数据更新到缓存中
            AyUser ayUser3 = ayUserService.findById("4");
            System.out.println("--->>> id: " + ayUser3.getId() + " name:" + ayUser3.getName());
            redisUserSize = redisTemplate.opsForList().size("ALL_USER_LIST");
            System.out.println("目前缓存中的用户数量为: " + redisUserSize);

    }
```

代码开发完成之后，重新启动项目，数据库中的 3 条数据会重新被添加到 Redis 缓存中，具体如图 7-12 所示。项目启动成功之后，我们往数据库表 ay_user 中添加 id 为 4 的第 4 条数据，具体如图 7-13 所示。

图 7-12　数据库钟存在 3 条数据

图 7-13　插入 id 为 4 的数据

最后执行单元测试方法 testFindById()，在 Intellij IDEA 的控制台中会打印如下的信息：

```
目前缓存中的用户数量为: 3
--->>> id: 1 name:阿毅
目前缓存中的用户数量为: 3
--->>> id: 2 name:阿兰
--->>> id: 4 name:test
目前缓存中的用户数量为: 4
```

第 8 章

集成 Log4J 日志

本章主要回顾 Log4J 的基础知识、在 Spring Boot 中集成 Log4J、Log4J 在 Spring Boot 中的运用以及如何把日志打印到控制台和记录到日志文件中等内容。

8.1 Log4J 概述

Log4J 是 Apache 下的一个开源项目,通过使用 Log4J 可以将日志信息打印到控制台、文件等,也可以控制每一条日志的输出格式,通过定义每一条日志信息的级别,更加细致地控制日志的生成过程。

在应用程序中添加日志记录有 3 个目的:

(1)监视代码中变量的变化情况,周期性地记录到文件中供其他应用进行统计分析工作。

(2)跟踪代码运行时轨迹,作为日后审计的依据。

(3)担当集成开发环境中的调试器的作用,向文件或控制台打印代码的调试信息。

Log4J 中有 3 个主要的组件，它们分别是：Logger（记录器）、Appender（输出源）和 Layout（布局），这 3 个组件可以简单地理解为日志类别、日志要输出的地方和日志以何种形式输出。Log4J 的原理如图 8-1 所示。

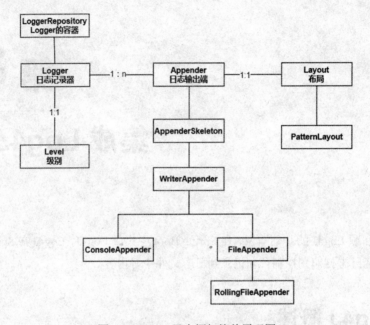

图 8-1　Log4J 日志框架简单原理图

- Logger（记录器）：Logger组件被分为7个级别：all、debug、info、warn、error、fatal、off。这7个级别是有优先级的，即：all<debug< info< warn< error< fatal<off，分别用来指定这条日志信息的重要程度。Log4J有一个规则——只输出级别不低于设定级别的日志信息。假设Logger级别设定为info，则info、warn、error和fatal级别的日志信息都会输出，而级别比info低的debug则不会输出。Log4J允许开发人员定义多个Logger，每个Logger拥有自己的名字，Logger之间通过名字来表明隶属关系。
- Appender（输出源）：Log4J日志系统允许把日志输出到不同的地方，如控制台（Console）、文件（Files）等，可以根据天数或者文件大小产生新的文件，可以以流的形式发送到其他地方等。
- Layout（布局）：Layout的作用是控制Log信息的输出方式，也就是格式化输出的信息。

Log4J 支持两种配置文件格式：一种是 XML 格式的文件；另一种是 Java 特性文件 log4J2.properties（键 = 值）。XML 文件可以配置更多的功能（如过滤），properties 文件简单易读，没有好坏，能够融会贯通就是最好的。具体的 XML 配置如下：

```xml
<?xml version="1.0" encoding="UTF-8"?>
<Configuration status="WARN">
    <Appenders>
        <Console name="Console" target="SYSTEM_OUT">
            <PatternLayout pattern="%d{HH:mm:ss.SSS} [%t] %-5level %logger{36} - %msg%n" />
        </Console>
    </Appenders>
    <Loggers>
        <Root level="debug">
            <AppenderRef ref="Console" />
        </Root>
    </Loggers>
</Configuration>
```

8.2 集成 Log4J2

8.2.1 引入依赖

在 Spring Boot 中集成 Log4J2，首先需要在 pom.xml 文件中引入所需的依赖，具体代码如下：

```xml
<!-- log4j2 -->
<dependency>
    <groupId>org.springframework.boot</groupId>
    <artifactId>spring-boot-starter-log4j2</artifactId>
</dependency>
```

Spring Boot 默认使用 Logback 日志框架来记录日志，并用 INFO 级别输出到控制台，所以我们在引入 Log4J2 之前，需要先排除该包的依赖，再引入 Log4J2 的依赖。具体做法是，找到 pom.xml 文件中的 spring-boot-starter-web 依赖，使用 exclusion 标签排除 Logback，具体排除 Logback 依赖的代码如下：

```xml
<dependency>
    <groupId>org.springframework.boot</groupId>
    <artifactId>spring-boot-starter-web</artifactId>
```

```xml
        <exclusions>
            <!-- 排查 Spring Boot 默认日志 -->
            <exclusion>
                <groupId>org.springframework.boot</groupId>
                <artifactId>spring-boot-starter-logging</artifactId>
            </exclusion>
        </exclusions>
    </dependency>
```

8.2.2 添加 Log4J 配置

在 8.1 节中已经讲过，Log4J2 支持两种配置文件格式：一种是 XML 格式的文件；另一种是 properties 文件的格式。这里使用 XML 格式配置 Log4J2，properties 格式可以作为大家的自学任务。使用 XML 格式配置很简单，只需要在 application.properties 文件中添加如下的配置信息：

```
###Log4J配置
logging.config=classpath:log4j2.xml
```

配置完成之后，Spring Boot 就会帮我们在 classpath 路径下查找 log4j2.xml 文件，所以最后一步，只需要配置好 log4j2.xml 文件即可。

8.2.3 创建 log4j2.xml 文件

application.properties 配置完成之后，在目录 /src/main/resources 下新建空的日志配置文件 log4j2.xml。具体代码如下：

```xml
<?xml version="1.0" encoding="UTF-8"?>
<Configuration status="WARN">
    <appenders>

    </appenders>
    <loggers>
        <root level="all">

        </root>
    </loggers>
</Configuration>
```

8.3 使用 Log4J 记录日志

8.3.1 打印到控制台

现在我们需要把日志打印到控制台，需要往 log4j2.xml 配置文件添加相关的配置，具体代码如下：

```xml
<?xml version="1.0" encoding="UTF-8"?>
<Configuration status="WARN">
    <appenders>
        <Console name="Console" target="SYSTEM_OUT">
            <!-- 指定日志的输出格式 -->
            <PatternLayout pattern="[%d{HH:mm:ss:SSS}] [%p] - %l - %m%n"/>
        </Console>
    </appenders>
    <loggers>
        <root level="info">
            <!-- 控制台输出 -->
            <appender-ref ref="Console"/>
        </root>
    </loggers>
</Configuration>
```

- <Console/>：指定控制台输出。
- <PatternLayout/>：指定日志的输出格式。

Spring Boot 集成 Log4J 日志完成之后，在 6.2.2 节和 7.3.1 节中，我们已经开发好 AyUserListener 监听器，但是使用 System.out.println 来打印信息是一种非常不合理的方式，现在我们把 Logger 类引入到 AyUserListener.java 监听器中，同时把 System.out.println 相关代码注释掉，改成用日志方式记录信息。这样，在项目启动过程中，调用上下文初始化和销毁方法的时候，就会记录日志到开发工具控制台或者日志文件中。AyUserListener 具体代码如下：

```
/**
 * 描述：监听器
 * @author Ay
```

```java
 * @date    2017/11/4
 */
@WebListener
public class AyUserListener implements ServletContextListener {
    //省略代码
    //需要添加的代码
    Logger logger = LogManager.getLogger(this.getClass());
    @Override
    public void contextInitialized(ServletContextEvent servletContextEvent) {
        //查询数据库所有的用户
        List<AyUser> ayUserList = ayUserService.findAll();
        //清除缓存中的用户数据
        redisTemplate.delete(ALL_USER);
        //存放到 Redis 缓存中
        redisTemplate.opsForList().leftPushAll(ALL_USER, ayUserList);
        //真实项目中需要注释掉
        List<AyUser> queryUserList = redisTemplate.opsForList().range(ALL_USER, 0, -1);
        //System.out.println("缓存中目前的用户数有: " + queryUserList.size() + " 人");
        //System.out.println("ServletContext 上下文初始化");
        logger.info("ServletContext 上下文初始化");
        logger.info("缓存中目前的用户数有: " + queryUserList.size() + " 人");
    }

    @Override
    public void contextDestroyed(ServletContextEvent servletContextEvent) {
        //System.out.println("ServletContext 上下文销毁");
        logger.info("ServletContext 上下文销毁");
    }
}
```

8.3.2 记录到文件

在 8.3.1 节中，日志只是被打印到控制台中，当项目真正被上线之后，是没有控制台这个概念的，上线环境中，项目的日志都是被记录到文件中的。我们继续在 log4j2.xml 配置文件中添加相关配置，使日志可以被打印到文件中，具体代码如下：

```xml
<?xml version="1.0" encoding="UTF-8"?>
<Configuration status="WARN">
    <appenders>
        <Console name="Console" target="SYSTEM_OUT">
            <!-- 设置日志输出的格式 -->
            <PatternLayout pattern="[%d{HH:mm:ss:SSS}] [%p] - %l - %m%n"/>
        </Console>
        <RollingFile name="RollingFileInfo" fileName="D:/info.log"
                filePattern="D:/$${date:yyyy-MM}/info-%d{yyyy-MM-dd}-%i.log">
            <Filters>
                <ThresholdFilter level="INFO"/>
            </Filters>
            <PatternLayout pattern="[%d{HH:mm:ss:SSS}] [%p] - %l - %m%n"/>
            <Policies>
                <TimeBasedTriggeringPolicy/>
                <SizeBasedTriggeringPolicy size="100 MB"/>
            </Policies>
        </RollingFile>
    </appenders>

    <loggers>
        <root level="info">
            <appender-ref ref="Console"/>
            <appender-ref ref="RollingFileInfo"/>
        </root>
    </loggers>

</Configuration>
```

- \<RollingFile\>标签：fileName用于定义日志的数据路径，如D:/info.log。filePattern定义日志的匹配方式。
- \<Filters\>标签：日志过滤策略，\<ThresholdFilter\>标签用于指定日志信息的最低输出级别，默认为DEBUG。

现在修改 3.2.3 节当中 AyUserServiceImpl 类的删除方法 Delete，我们希望删除用户这个操作可以被记录到日志文件中，AyUserServiceImpl 类代码具体的修改如下：

```java
/**
 * 描述：用户服务层实现类
 * @author 阿毅
 * @date   2017/10/14
 */
//@Transactional
@Service
public class AyUserServiceImpl implements AyUserService {
    //省略代码
    //需要添加的代码
    Logger logger = LogManager.getLogger(this.getClass());

    @Override
    public void delete(String id) {
        ayUserRepository.delete(id);
        //需要添加的代码
        logger.info("userId:" + id + "用户被删除");
    }

    //省略代码
}
```

8.3.3 测试

代码开发完成之后，接下来就是测试工作了。重新启动项目，启动之前，记得打开 Redis 服务器，因为之前的章节已经在 Spring Boot 中整合了 Redis，项目重启的过程中，可以在 Intellij IDEA 控制台中看到如图 8-2 所示的信息。同时，我们可以到 D 盘查看日志文件 info.log，在日志文件中按住 Ctrl + F 键，可以查询到和图 8-2 所示一样的信息。

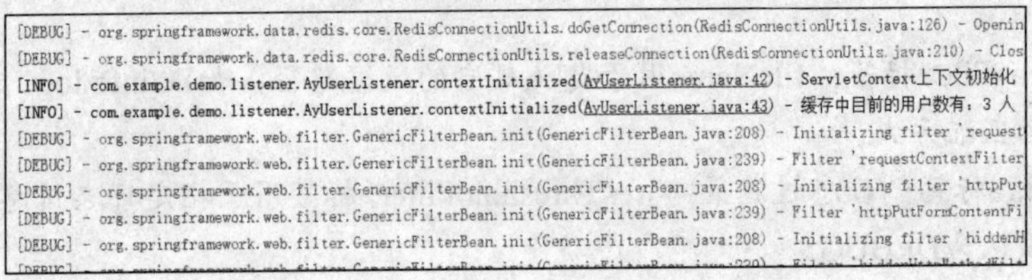

图 8-2 Redis 断点调试界面

接着，再测试一下删除用户的时候，日志是否可以打印到控制台或者记录到日志文件中。在测试类 DemoApplicationTests 下添加测试用例，具体代码如下：

```
@RunWith(SpringRunner.class)
@SpringBootTest
public class DemoApplicationTests{
    //省略代码
    Logger logger = LogManager.getLogger(this.getClass());

    @Test
    public void testLog4j(){
        ayUserService.delete("4");
        logger.info("delete success!!!");
    }
}
```

在数据库 ay_test 表中存在 4 条数据，如图 8-3 所示。运行单元测试方法 testLog4J，如果同样可以在 Intellij IDEA 控制台或者 D 盘的 info.log 文件中打印如图 8-4 所示信息，证明 Spring Boot 整合 Log4J 以及在 Spring Boot 中运用 Log4J 成功。

图 8-3 Redis 断点调试界面　　　　图 8-4 Redis 断点调试界面

提示　启动项目的时候，记得启动 Redis 服务器，否则会报错。

8.4 思考题

Spring Boot 支持哪些日志框架？推荐和默认的日志框架是哪个？

答：Spring Boot 支持 Java Util Logging、Log4J2 及 Lockback 作为日志框架，如果你使用 Starters 启动器，Spring Boot 将使用 Logback 作为默认日志框架。

第 9 章

Quartz 定时器和发送 Email

本章主要介绍在 Spring Boot 中使用 XML 配置和 Java 注解两种方式定义和使用 Quartz 定时器，以及如何在 Spring Boot 中通过 JavaMailSender 接口给用户发送广告邮件等内容。

9.1 使用 Quartz 定时器

9.1.1 Quartz 概述

Quartz 是一个完全由 Java 编写的开源任务调度的框架，通过触发器设置作业定时运行规则，控制作业的运行时间。Quartz 定时器作用很多，比如定时发送信息、定时生成报表等。

Quartz 框架主要核心组件包括调度器、触发器、作业，调度器作为作业的总指挥，触发器作为作业的操作者，作业为应用的功能模块，其关系如图 9-1 所示。

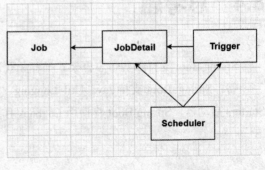

图 9-1 Quartz 各个组件的关系

Job 是一个接口，该接口只有一个方法 execute，被调度的作业（类）需实现该接口中的 execute() 方法，JobExecutionContext 类提供了调度上下文的各种信息。每次执行该 Job 均重新创建一个 Job 实例。Job 的源代码如下所示：

```
public interface Job {
    void execute(JobExecutionContext var1) throws JobExecutionException;
}
```

Quartz 在每次执行 Job 时，都重新创建一个 Job 实例，所以它不直接接收一个 Job 的实例，相反它接收一个 Job 实现类，以便运行时通过 newInstance() 的反射机制实例化 Job。因此需要通过一个类来描述 Job 的实现类及其他相关的静态信息，如 Job 名字、描述、关联监听器等信息，JobDetail 承担了这一角色。JobDetail 用来保存作业的详细信息。一个 JobDetail 可以有多个 Trigger，但是一个 Trigger 只能对应一个 JobDetail。

Trigger 触发器描述触发 Job 的执行规则。主要有 SimpleTrigger 和 CronTrigger 这两个子类。当仅需触发一次或者以固定时间间隔周期执行时，SimpleTrigger 是最适合的选择；而 CronTrigger 则可以通过 Cron 表达式定义出各种复杂时间规则的调度方案：如每天早晨 9:00 执行，周一、周三、周五下午 5:00 执行等。

CronTrigger 配置格式：

格式：	[秒] [分] [小时] [日] [月] [周] [年]
0 0 12 * * ?	每天 12 点触发
0 15 10 ? * *	每天 10 点 15 分触发
0 15 10 * * ?	每天 10 点 15 分触发
0 15 10 * * ? *	每天 10 点 15 分触发
0 15 10 * * ? 2005	2005 年每天 10 点 15 分触发
0 * 14 * * ?	每天下午的 2 点到 2 点 59 分每分触发
0 0/5 14 * * ?	每天下午的 2 点到 2 点 59 分 (整点开始，每隔 5 分触发)
0 0/5 14,18 * * ?	每天下午的 18 点到 18 点 59 分 (整点开始，每隔 5 分触发)
0 0-5 14 * * ?	每天下午的 2 点到 2 点 05 分每分触发
0 10,44 14 ? 3 WED	3 月份每周三下午的 2 点 10 分和 2 点 44 分触发
0 15 10 ? * MON-FRI	从周一到周五每天上午的 10 点 15 分触发
0 15 10 15 * ?	每月 15 号上午 10 点 15 分触发
0 15 10 L * ?	每月最后一天的 10 点 15 分触发
0 15 10 ? * 6L	每月最后一周的星期五的 10 点 15 分触发
0 15 10 ? * 6L 2002-2005	从 2002 年到 2005 年每月最后一周的星期五的 10 点 15 分触发
0 15 10 ? * 6#3	每月的第三周的星期五开始触发
0 0 12 1/5 * ?	每月的第一个中午开始每隔 5 天触发一次
0 11 11 11 11 ?	每年的 11 月 11 号 11 点 11 分触发 (光棍节)

Scheduler 负责管理 Quartz 的运行环境，Quartz 是基于多线程架构的，它启动的时候会初始化一套线程，这套线程会用来执行一些预置的作业。Trigger 和 JobDetail 可以注册到 Scheduler 中。Scheduler 可以将 Trigger 绑定到某一个 JobDetail 中，这样当 Trigger 触发时，对应的 Job 就被执行。Scheduler 拥有一个 SchedulerContext，它类似于 ServletContext，保存着 Scheduler 上下文信息，Job 和 Trigger 都可以访问 SchedulerContext 内的信息。Scheduler 使用一个线程池作为任务运行的基础设施，任务通过共享线程池中的线程提高运行效率。

9.1.2 引入依赖

在 Spring Boot 中集成 Quartz，首先需要在 pom.xml 文件中引入所需的依赖，具体代码如下：

```xml
<!-- quartz 定时器 -->
<dependency>
    <groupId>org.quartz-scheduler</groupId>
    <artifactId>quartz</artifactId>
    <version>2.2.3</version>
</dependency>
```

9.1.3 定时器配置文件

创建定时器的方法有两种：① 使用 XML 配置文件的方式；② 使用注解的方式。注解的方式不需要任何配置文件且简单高效，这两种方式都会讲到。我们先来讲第一种方式，也就是配置文件的方式。首先，我们需要在/src/main/resources 目录下新建配置文件 spring-mvc.xml，具体代码如下：

```xml
<beans xmlns="http://www.springframework.org/schema/beans"
    xmlns:xsi="http://www.w3.org/2001/XMLSchema-instance"
    xmlns:context="http://www.springframework.org/schema/context"
    xmlns:mvc="http://www.springframework.org/schema/mvc"
    xmlns:aop="http://www.springframework.org/schema/aop"
    xmlns:task="http://www.springframework.org/schema/task"
    xsi:schemaLocation="
    http://www.springframework.org/schema/beans
    http://www.springframework.org/schema/beans/spring-beans.xsd
    http://www.springframework.org/schema/context
    http://www.springframework.org/schema/context/spring-context-4.2.xsd
    http://www.springframework.org/schema/mvc
```

```xml
        http://www.springframework.org/schema/mvc/spring-mvc.xsd
        http://www.springframework.org/schema/aop
        http://www.springframework.org/schema/aop/spring-aop.xsd
        http://www.springframework.org/schema/task
        http://www.springframework.org/schema/task/spring-task.xsd">

    <context:annotation-config/>
    <!-- 利用import引入定时器的文件 -->
    <import resource="spring-quartz.xml"/>

</beans>
```

- `<import>`标签：import标签用于导入定时器的配置文件，该标签可以根据具体业务分离配置文件。

然后，我们在/src/main/resources 目录下新建 spring-quartz.xml 配置文件，具体代码如下：

```xml
<beans xmlns="http://www.springframework.org/schema/beans"
       xmlns:xsi="http://www.w3.org/2001/XMLSchema-instance"
       xsi:schemaLocation="http://www.springframework.org/schema/beans
       http://www.springframework.org/schema/beans/spring-beans-3.0.xsd">

<!-- 定义Job对象 -->
<bean id="taskJob" class="com.example.demo.quartz.TestTask"/>
    <!-- 定义JobDetail对象 -->
    <bean id="JobDetail"
    class="org.springframework.scheduling.quartz.MethodInvokingJobDetailFactoryBean">
    <!-- 目标对象 taskJob -->
    <property name="targetObject">
        <ref bean="taskJob"/>
    </property>
    <!-- 目标方法 -->
        <property name="targetMethod">
            <value>run</value>
        </property>
    </bean>

    <!-- 调度触发器 -->
```

```xml
    <bean id="myTrigger"
        class="org.springframework.scheduling.quartz.CronTriggerFactoryBean">
        <!-- 指定使用 jobDetail -->
        <property name="jobDetail">
            <ref bean="jobDetail" />
        </property>
        <!-- 定义触发规则，每 10 秒执行一次 -->
        <property name="cronExpression">
            <value>0/10 * * * * ?</value>
        </property>
    </bean>


    <!-- 调度工厂 -->
    <bean id="scheduler"
        class="org.springframework.scheduling.quartz.SchedulerFactoryBean">
        <!-- 注册触发器，可注册多个 -->
        <property name="triggers">
            <list>
                <ref bean="myTrigger"/>
            </list>
        </property>
    </bean>
</beans>
```

在 spring-quartz.xml 配置文件中，我们分别定义了 Job、JobDetail、Trigger 以及 Scheduler。并配置了它们之间的关系。

9.1.4　创建定时器类

定时器的依赖及配置文件开发完成之后，在 /src/main/java/com.example.demo.quartz 目录下新建定时器类 TestTask.java。具体代码如下：

```java
/**
 * 描述: 定时器类
 * @author Ay
 * @date   2017/11/18
 */
```

```java
public class TestTask {

    //日志对象
    private static final Logger logger = LogManager.getLogger(TestTask.class);

    public void run() {
        logger.info("定时器运行了!!!");
    }
}
```

如果使用第二种创建定时器的方法，相对就简单了，只需要创建一个定时器类，加上相关的注解就搞定了。比如，我们可以在 /src/main/java/com.example.demo.quartz 目录下创建 SendMailQuartz 定时器类，具体代码如下：

```java
/**
 * 描述：定时器类
 * @author Ay
 * @date   2017/11/18
 */
@Component
@Configurable
@EnableScheduling
public class SendMailQuartz {

    //日志对象
    private static final Logger logger = LogManager.getLogger(SendMailQuartz.class);

    //每5秒执行一次
    @Scheduled(cron = "*/5 * * * * ")
    public void reportCurrentByCron(){
        logger.info("定时器运行了!!!");
    }

}
```

- @Configurable：加上此注解的类相当于XML配置文件可以被Spring Boot扫描初始化。
- @EnableScheduling：通过在配置类注解@EnableScheduling来开启对计划任务的支持，然后在要执行计划任务的方法上注解@Scheduled，声明这是一个计划任务。

- @Scheduled：注解为定时任务，cron表达式里写执行的时机。

9.1.5　Spring Boot 扫描配置文件

在 9.1.3 节中已经开发好 spring-mvc.xml 配置文件，但是想让 Spring Boot 扫描到该配置文件，还需要在入口类 MySpringBootApplication 中添加@ImportResource 注解，具体代码如下：

```
@SpringBootApplication
@ServletComponentScan
@ImportResource(locations={"classpath:spring-mvc.xml"})
public class MySpringBootApplication {
    //省略代码
}
```

- @ImportResource：导入资源配置文件，让 Spring Boot可以读取到，类似于XML配置文件中的<import>标签。

9.1.6　测试

代码开发完成之后，重新启动项目，在 Intellij IDEA 控制台中可以看到如图 9-2 所示的信息，证明在 Spring Boot 中整合 Quartz 定时器成功。

```
org.springframework.transaction.support.TransactionSynchronizationManager.doUnbindResource(TransactionSynchro
org.quartz.core.JobRunShell.run(JobRunShell.java:201) - Calling execute on job DEFAULT.jobDetail
org.quartz.core.QuartzSchedulerThread.run(QuartzSchedulerThread.java:276) - batch acquisition of 1 triggers
com.example.demo.quartz.TestTask.run(TestTask.java:17) - 定时器运行了!!!
com.example.demo.quartz.SendMailQuartz.reportCurrentByCron(SendMailQuartz.java:40) - 定时器运行了!!!
org.springframework.transaction.support.TransactionSynchronizationManager.bindResource(TransactionSynchroniza
org.springframework.transaction.support.AbstractPlatformTransactionManager.getTransaction(AbstractPlatformTra
```

图 9-2　Quartz 定时器打印信息

9.2　Spring Boot 发送 Email

9.2.1　Email 介绍

邮件服务在互联网早期就已经出现，如今已成为人们互联网生活中必不可少的一项服务。邮件发送与接收的过程如下：

（1）发件人使用 SMTP 协议传输邮件到邮件服务器 A。
（2）邮件服务器 A 根据邮件中指定的接收者，投送邮件至相应的邮件服务器 B。
（3）收件人使用 POP3 协议从邮件服务器 B 接收邮件。

SMTP（Simple Mail Transfer Protocol）是电子邮件（Email）传输的互联网标准，定义在 RFC5321，默认使用端口 25。

POP3（Post Office Protocol - Version 3）主要用于支持使用客户端远程管理在服务器上的电子邮件。定义在 RFC 1939，为 POP 协议的第三版（最新版）。

这两个协议均属于 TCP/IP 协议族的应用层协议，运行在 TCP 层之上。

发送邮件的需求比较常见，如找回密码、事件通知、向用户发送广告邮件等。Sun 公司给广大 Java 开发人员提供了一款邮件发送和接收的开源类库 JavaMail，支持常用的邮件协议，如 SMTP、POP3、IMAP 等。开发人员使用 JavaMail 编写邮件程序时，不再需要考虑底层的通信细节（如 Socket），而是关注逻辑层面。JavaMail 可以发送各种复杂的 MIME 格式的邮件内容，注意 JavaMail 仅支持 JDK4 及以上版本。虽然 JavaMail 是 JDK 的 API，但它并没有直接加入 JDK 中，所以我们需要另外添加依赖。

Spring 提供了非常好用的 JavaMailSender 接口实现邮件发送，在 Spring Boot 的 Starter 模块中已为此提供了自动化配置。

9.2.2　引入依赖

在 Spring Boot 中集成 Mail，首先需要在 pom.xml 文件中引入所需的依赖，具体代码如下：

```
<!-- mail start -->
<dependency>
    <groupId>org.springframework.boot</groupId>
    <artifactId>spring-boot-starter-mail</artifactId>
</dependency>
```

9.2.3　添加 Email 配置

在 pom 文件引入 Mail 所需的依赖之后，需要在 application.properties 文件中添加如下的配置信息：

```
###Mail 邮件配置
###邮箱主机
```

```
spring.mail.host=smtp.163.com
###用户名
spring.mail.username=huangwenyi10@163.com
###设置的授权码
spring.mail.password=自己邮箱密码
###默认编码
spring.mail.default-encoding=UTF-8
spring.mail.properties.mail.smtp.auth=true
spring.mail.properties.mail.smtp.starttls.enable=true
spring.mail.properties.mail.smtp.starttls.required=true
```

9.2.4 在定时器中发送邮件

在 Spring Boot 中添加完依赖和配置之后，在项目的/src/main/java/com.example.demo.mail 目录下新建邮件服务接口类 SendJunkMailService，具体代码如下：

```java
/**
 * 描述: 发送用户邮件服务
 * @author Ay
 * @date   2017/11/19
 */
public interface SendJunkMailService {

    boolean sendJunkMail(List<AyUser> ayUser);
}
```

然后，继续在项目的目录/src/main/java/com.example.demo.mail.impl 下新建接口类的实现类 SendJunkMailServiceImpl.java，具体代码如下：

```java
/**
 * 描述: 发送用户邮件服务
 * @author Ay
 * @date   2017/11/19
 */
@Service
public class SendJunkMailServiceImpl implements SendJunkMailService {

    @Autowired
    JavaMailSender mailSender;
```

```java
        @Resource
        private AyUserService ayUserService;
        @Value("${spring.mail.username}")
        private String from;
        public static final Logger logger = LogManager.getLogger
(SendJunkMailServiceImpl.class);

        @Override
        public boolean sendJunkMail(List<AyUser> ayUserList) {

            try{
                if(ayUserList == null || ayUserList.size() <= 0 ) return Boolean.FALSE;
                for(AyUser ayUser: ayUserList){
                    MimeMessage mimeMessage = this.mailSender.createMimeMessage();
                    MimeMessageHelper message = new MimeMessageHelper(mimeMessage);
                    //邮件发送方
                    message.setFrom(from);
                    //邮件主题
                    message.setSubject("地瓜今日特卖");
                    //邮件接收方
                    message.setTo("al_test@163.com");
                    //邮件内容
                    message.setText(ayUser.getName() +" ,你知道么？厦门地瓜今日特卖，一斤只要9元");
                    //发送邮件
                    this.mailSender.send(mimeMessage);
                }
            }catch(Exception ex){
                logger.error("sendJunkMail error and ayUser=%s", ayUserList, ex);
                return Boolean.FALSE;
            }
            return Boolean.TRUE;
        }
    }
```

- @Value：可以将application.properties配置文件中的配置设置到属性中。如上面代码中，会将spring.mail.username的值huangwenyi10@163.com设置给from属性。

- **JavaMailSender**：邮件发送接口。在 Spring Boot 的 Starter 模块中已为此提供了自动化配置。我们只需要通过注解 @Autowired 注入进来，即可使用。

在 9.1.4 节中已经开发了 SendMailQuartz 定时器类，现在重新修改该类，让定时器类能够每隔一段时间给数据库的用户发送广告邮件，SendMailQuartz 类具体的修改如下：

```java
/**
 * 描述：定时器类
 * @author Ay
 * @date   2017/11/18
 */
@Component
@Configurable
@EnableScheduling
public class SendMailQuartz {

    //日志对象
    private static final Logger logger = LogManager.getLogger(SendMailQuartz.class);

    @Resource
    private SendJunkMailService sendJunkMailService;
    @Resource
    private AyUserService ayUserService;

    //每5秒执行一次
    @Scheduled(cron = "*/5 * * * * ")
    public void reportCurrentByCron(){
        List<AyUser> userList = ayUserService.findAll();
        if (userList == null || userList.size() <= 0) return;
        //发送邮件
        sendJunkMailService.sendJunkMail(userList);
        logger.info("定时器运行了!!!");
    }

}
```

9.2.5 测试

代码全部开发完成之后，重新启动项目，发送邮件定时器类 SendMailQuartz 每隔 5 秒（真实项目会设置比较长，比如 1 天、2 天等）会查询数据库表 ay_test 中的所有用户，并发送广告邮件给对应的用户。我们登录 al_test@163.com 邮箱，便可以查看到如图 9-3 和图 9-4 所示的信息。

图 9-3　163 邮箱界面　　　　　　　图 9-4　163 邮件内容

第 10 章

集成 MyBatis

本章主要介绍如何在 Spring Boot 中集成 MyBatis 框架,以及通过 MyBatis 框架实现查询等功能,最后介绍如何使用 MyBatisCodeHelper 插件快速生成增删改查代码等内容。

10.1 MyBatis 介绍

MyBatis 是一款优秀的持久层框架,它支持定制化 SQL、存储过程以及高级映射。MyBatis 避免了几乎所有的 JDBC 代码和手动设置参数以及获取结果集。MyBatis 可以使用简单的 XML 或注解来配置和映射原生信息,将接口和 Java 的 POJOs(Plain Old Java Objects,普通的 Java 对象)映射成数据库中的记录。

10.2 集成 MyBatis 的步骤

10.2.1 引入依赖

在 Spring Boot 中集成 MyBatis,首先需要在 pom.xml 文件中引入所需的依赖,具体代码如下:

```xml
<!-- mybatis start -->
    <dependency>
        <groupId>org.mybatis.spring.boot</groupId>
        <artifactId>mybatis-spring-boot-starter</artifactId>
        <version>2.0.1</version>
    </dependency>
```

10.2.2　添加 MyBatis 配置

在 pom 文件添加 MyBatis 所需的依赖之后，需要在 application.properties 文件中添加如下的配置信息：

```
### MyBatis 配置
mybatis.mapper-locations=classpath:/mappers/*Mapper.xml
mybatis.type-aliases-package=com.example.demo.dao
```

- mybatis.mapper-locations：Mapper 资源文件存放的路径。
- mybatis.type-aliases-package：Dao 接口文件存放的目录。

10.2.3　Dao 层和 Mapper 文件开发

application.properties 配置添加完成之后，需要根据 MyBatis 配置添加对应的文件夹。首先，需要在 /src/main/java/com.example.demo.dao 目录下新建 AyUserDao 接口，这样 Spring Boot 启动时，就可以根据 application.properties 配置 mybatis.type-aliases-package，找到 AyUserDao 接口。AyUserDao 的具体代码如下：

```java
/**
 * 描述：用户 DAO
 * @author Ay
 * @date   2017/11/20
 */
@Mapper
public interface AyUserDao {

    /**
     * 描述：通过用户名和密码查询用户
     * @param name
```

```
     * @param password
     */
    AyUser findByNameAndPassword(@Param("name") String name,
@Param("password") String password);

}
```

- @Mapper：重要注解，MyBatis根据接口定义与Mapper文件中的SQL语句动态创建接口实现。
- @Param：注解参数，在Mapper.xml配置文件中，可以采用#{}的方式对@Param注解括号内的参数进行引用。
- findByNameAndPassword：该方法可以通过用户名和密码查询用户。

然后，在/src/main/resources/mappers 目录下新建 AyUserMapper.xml 文件，Spring Boot 在项目启动时，会根据 application.properties 配置 mybatis.mapper-locations 找到该文件。AyUserMapper.xml 的具体代码如下：

```xml
<?xml version="1.0" encoding="UTF-8" ?>
<!DOCTYPE mapper PUBLIC "-//mybatis.org//DTD Mapper 3.0//EN"
        "http://mybatis.org/dtd/mybatis-3-mapper.dtd" >
<mapper namespace="com.example.demo.dao.AyUserDao" >

    <resultMap id="UserResultMap" type="com.example.demo.model.AyUser">
        <id column="id" property="id" jdbcType="VARCHAR"/>
        <result column="name" property="name" jdbcType="VARCHAR"/>
        <result column="password" property="password" jdbcType="VARCHAR"/>
        <result column="mail" property="mail" jdbcType="VARCHAR"/>
    </resultMap>

    <select id="findByNameAndPassword" resultMap="UserResultMap" parameterType="String">
        select * from ay_user u
        <where>
           u.name = #{name}
           and u.password = #{password}
        </where>
```

```
        </select>

</mapper>
```

- `<mapper>`标签：该标签的namespace属性用于绑定Dao接口。
- `<select>`标签：用来编写select语句，映射查询语句。select标签有几个重要的属性，比如resultMap。
- `<resultMap>`：映射管理器resultMap，是MyBatis中最强大的工具，描述了如何将数据库查询的结果集映射到Java对象，并管理结果和实体类之间的映射关系。

AyUserMapper.xml 类开发完成之后，需要在目录/src/main/java/com.example.demo.model 下开发对应的实体类 AyUser，具体代码如下：

```
/**
 * @author Ay
 * @create 2019/07/02
 **/
@Entity
@Table(name = "ay_user")
public class AyUser implements Serializable {

    //主键
    @Id
    private String id;
    //用户名
    private String name;
    //密码
    private String password;
    //邮箱
    private String mail;

    //省略set、get方法
}
```

AyUser 代码开发完成之后，在之前开发好的 AyUserService 接口类中添加接口 findByNameAndPassword。具体代码如下：

```java
/**
 * 描述：用户服务层接口
 * @author 阿毅
 * @date   2017/10/14
 */
public interface AyUserService {
    //此处省略代码

    AyUser findByNameAndPassword(String name, String password);

}
```

然后，在 AyUserServiceImpl 类中实现 findByNameAndPassword 接口，具体代码如下：

```java
/**
 * 描述：用户服务层实现类
 * @author 阿毅
 * @date   2017/10/14
 */
//@Transactional
@Service
public class AyUserServiceImpl implements AyUserService {

    //此处省略代码

    @Resource
    private AyUserDao ayUserDao;

    @Override
    public AyUser findByNameAndPassword(String name, String password) {
        return ayUserDao.findByNameAndPassword(name, password);
    }
}
```

10.2.4 测试

代码开发完成之后，在 MySpringBootApplicationTests 类下添加测试方法，具体代码如下：

```
@Resource
private AyUserService ayUserService;

@Test
public void testMybatis(){
    AyUser ayUser = ayUserService.findByNameAndPassword("阿毅", "123456");
    logger.info(ayUser.getId() + ayUser.getName());

}
```

执行测试用例，在 Intellij IDEA 控制台上可以看到相应的打印信息。

第 11 章

异步消息与异步调用

本章主要介绍 ActiveMQ 的安装与使用、Spring Boot 集成 ActiveMQ、利用 ActiveMQ 实现异步发表微信说说以及 Spring Boot 异步调用@Async 等内容。

11.1　JMS 消息概述

JMS（Java Message Service，即 Java 消息服务）是一组 Java 应用程序接口，它提供消息的创建、发送、读取等一系列服务。JMS 提供了一组公共应用程序接口和响应的语法，类似于 Java 数据库的统一访问接口 JDBC，它是一种与厂商无关的 API，使得 Java 程序能够与不同厂商的消息组件很好地进行通信。

JMS 支持两种消息发送和接收模型。一种称为 P2P（Ponit to Point）模型，即采用点对点的方式发送消息。P2P 模型是基于队列的，消息生产者（Producer）发送消息到队列，消息消费者（Consumer）从队列中接收消息，队列的存在使消息的异步传输成为可能。P2P 模式图如图 11-1 所示。

P2P 的特点是每个消息只有一个消费者（一旦被消费，消息就不在消息队列中），发送者和接收者之间在时间上没有依赖性，也就是说当发送者发送了消息之后，不管接收者有没有正在运行，它不会影响消息被发送到队列中，接收者在成功接收消息之后需向队列应答成功。

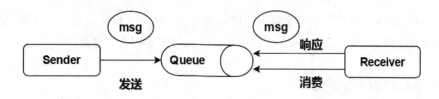

图 11-1　P2P 模式图

另一种称为 Pub/Sub（Publish/Subscribe，即发布-订阅）模型，发布-订阅模型定义了如何向一个内容节点发布和订阅消息，这个内容节点称为 Topic（主题）。主题可以认为是消息传递的中介，消息发布者将消息发布到某个主题，而消息订阅者则从主题订阅消息。主题使得消息的订阅者与消息的发布者互相保持独立，不需要进行接触即可保证消息的传递，发布-订阅模型在消息的一对多广播时采用。Pub/Sub 模式图如图 11-2 所示。

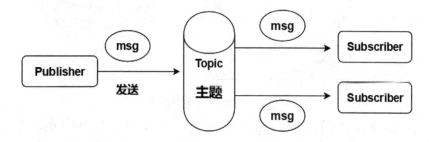

图 11-2　Pub/Sub 模式图

Pub/Sub 的特点是每个消息可以有多个消费者，发布者和订阅者之间有时间上的依赖性。针对某个主题（Topic）的订阅者，它必须创建一个订阅者之后，才能消费发布者的消息，而且为了消费消息，订阅者必须保持运行的状态。为了缓和这样严格的时间相关性，JMS 允许订阅者创建一个可持久化的订阅。这样，即使订阅者没有被激活（运行），它也能接收到发布者的消息。如果你希望发送的消息不被做任何处理，或者被一个消息者处理，或者可以被多个消费者处理的话，那么可以采用 Pub/Sub 模型。

11.2　Spring Boot 集成 ActiveMQ

11.2.1　ActiveMQ 概述

MQ 全称为 MessageQueue（消息队列），是一个消息的接收和转发的容器，可用于消息推送。ActiveMQ 是 Apache 提供的一个开源的消息系统，完全采用 Java 来实现，因此，它能很好地支持 J2EE 提出的 JMS（Java Message Service，即 Java 消息服务）规范。

11.2.2　ActiveMQ 的安装

安装 ActiveMQ 之前，需要到官方网站（http://activemq.apache.org/activemq-5150-release.html）下载，本书使用 apache-activemq-5.15.0 这个版本进行讲解。ActiveMQ 的具体安装步骤如下：

步骤 01　将官方网站下载的安装包 apache-activemq-5.15.0-bin.zip 解压。

步骤 02　打开解压的文件夹，进入 bin 目录，根据电脑操作系统是 32 位还是 64 位，选择进入【win32】文件夹或者【win64】文件夹，如图 11-3 所示。

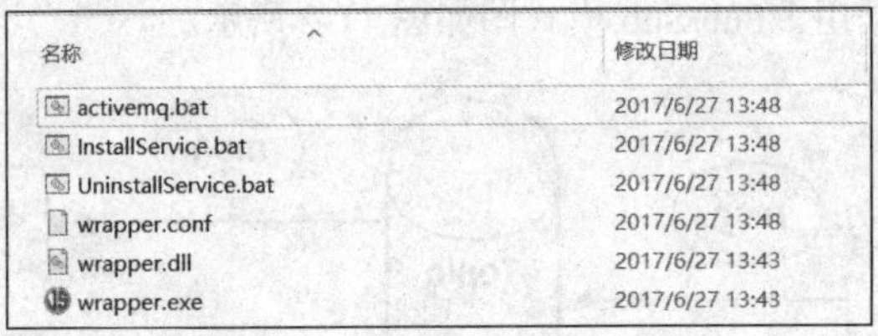

图 11-3　ActiveMQ 安装界面

步骤 03　双击【activemq.bat】，即可启动 ActiveMQ。当看到如图 11-4 所示的启动信息时，代表 ActiveMQ 安装成功。从图中可以看出，ActiveMQ 默认启动到 8161 端口。

图 11-4　ActiveMQ 启动成功界面

步骤 04　安装成功之后，在浏览器中输入 http://localhost:8161/admin 链接访问，第一次访问需要输入用户名 admin 和密码 admin 进行登录，登录成功之后，就可以看到 ActiveMQ 的首页。具体如图 11-5 所示。

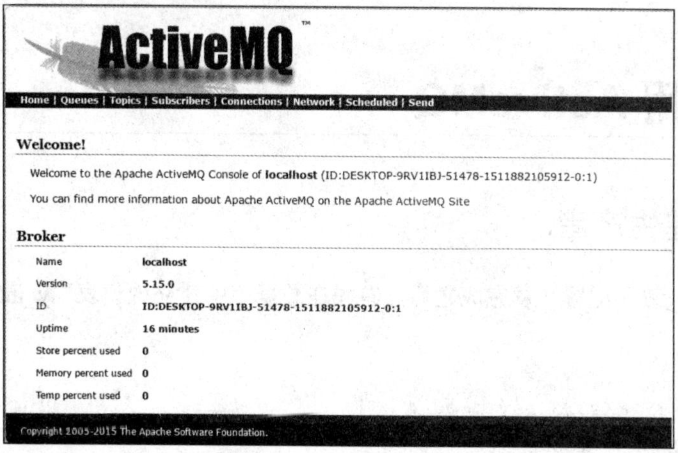

图 11-5　ActiveMQ 首页

11.2.3　引入依赖

在 Spring Boot 中集成 ActiveMQ，首先需要在 pom.xml 文件中引入所需的依赖，具体代码如下：

```xml
<!-- activemq start -->
<dependency>
    <groupId>org.springframework.boot</groupId>
    <artifactId>spring-boot-starter-activemq</artifactId>
</dependency>
```

11.2.4　添加 ActiveMQ 配置

依赖添加完成之后，需要在 application.properties 配置文件中添加 ActiveMQ 配置，具体代码如下：

```
spring.activemq.broker-url=tcp://localhost:61616
spring.activemq.in-memory=true
spring.activemq.pool.enabled=false
spring.activemq.packages.trust-all=true
```

- spring.activemq.packages.trust-all：ObjectMessage的使用机制是不安全的，ActiveMQ自5.12.2和5.13.0之后，强制Consumer端声明一份可信任的包列表，只有当ObjectMessage中的Object在可信任包内，才能被提取出来。该配置表示信任所有的包。

11.3 使用 ActiveMQ

11.3.1 创建生产者

ActiveMQ 依赖和配置开发完成之后,首先在数据库中建立说说表 ay_mood。具体建表语句如下:

```sql
DROP TABLE IF EXISTS `ay_mood`;
CREATE TABLE `ay_mood` (
  `id` varchar(32) NOT NULL,
  `content` varchar(256) DEFAULT NULL,
  `user_id` varchar(32) DEFAULT NULL,
  `praise_num` int(11) DEFAULT NULL,
  `publish_time` datetime DEFAULT NULL,
  PRIMARY KEY (`id`),
  KEY `mood_user_id_index` (`user_id`) USING BTREE
) ENGINE=InnoDB DEFAULT CHARSET=utf8;
```

数据库表建好之后,在 src\main\java\com\example\demo\model\ 目录下开发对应的 Java Bean 对象,具体代码如下:

```java
/**
 * 描述: 微信说说实体
 * @author Ay
 * @date   2017/11/28
 */
@Entity
@Table(name = "ay_mood")
public class AyMood implements Serializable {
    //主键
    @Id
    private String id;
    //说说内容
    private String content;
    //用户 id
    private String userId;
```

```
    //点赞数量
    private Integer praiseNum;
    //发表时间
    private Date publishTime;

    //省略 set、get 方法
}
```

AyMood 实体对象开发完成之后,开发对应的 AyMoodRepository 接口,具体代码如下:

```
/**
 * 描述: 说说 repository
 * @author Ay
 * @date   2017/12/02
 */
public interface AyMoodRepository extends JpaRepository<AyMood,String> {

}
```

Repository 接口开发完成之后,开发对应的说说服务层接口 AyMoodService 和相应的实现类 AyMoodServiceImpl。AyMoodService 的具体代码如下:

```
/**
 * 描述: 微信说说服务层
 * @author Ay
 * @date   2017/12/2
 */
public interface AyMoodService {

    AyMood save(AyMood ayMood);
}
```

AyMoodService 代码很简单,只有一个保存说说的方法 save(),AyMoodService 开发完成之后,实现该接口,具体代码如下:

```
/**
 * 描述: 微信说说服务层
 * @author Ay
 * @date   2017/12/2
 */
```

```
@Service
public class AyMoodServiceImpl implements AyMoodService{

    @Resource
    private AyMoodRepository ayMoodRepository;

    @Override
    public AyMood save(AyMood ayMood) {
        return ayMoodRepository.save(ayMood);
    }
}
```

在实现类 AyMoodServiceImpl 中注入 AyMoodRepository 接口,并调用其提供的 sava() 方法,保存说说到数据库。代码开发完成之后,在测试类 DemoApplicationTests 下添加测试方法:

```
@Resource
private AyMoodService ayMoodService;

@Test
public void testAyMood(){
    AyMood ayMood = new AyMood();
    ayMood.setId("1");
    //用户阿毅 id 为 1
    ayMood.setUserId("1");
    ayMood.setPraiseNum(0);
    //说说内容
    ayMood.setContent("这是我的第一条微信说说!!!");
    ayMood.setPublishTime(new Date());
    //往数据库保存一条数据,代码用户阿毅发表了一条说说
    AyMood mood = ayMoodService.save(ayMood);
}
```

测试用例 testAyMood 开发完成之后,我们允许它。运行成功之后,可以在数据库表 ay_mood 中看到一条记录,具体如图 11-6 所示。

id	content	user_id	praise_num	publish_time
1	这是我的第一条微信说说!!!	1	0	2019-07-15 15:08:27

图 11-6　说说表记录

用户发表说说的类虽然都开发完成了,但是有一个问题,我们知道微信的用户量极大,每天都有几亿的用户发表不同的说说,如果按照上面的做法,用户每发一条说说,后端都会单独开一个线程,将该说说的内容实时地保存到数据库中。由于后端服务系统的线程数和数据库线程池中的线程数量是有限而且宝贵的,将用户发表的说说实时保存到数据库中,必然造成后端服务和数据库极大的压力,所以需要使用 ActiveMQ 作异步消费,来减轻用户大并发表说说而产生的压力,提高系统的整体性能。

下面来开发生产者和消费者。生产者 AyMoodProducer 的代码如下:

```
/**
 * 生产者
 * @author Ay
 * @date 2017/11/30
 */
@Service
public class AyMoodProducer{

    @Resource
    private JmsMessagingTemplate jmsMessagingTemplate;

    public void sendMessage(Destination destination, final String message) {
        jmsMessagingTemplate.convertAndSend(destination, message);
    }

}
```

- JmsMessagingTemplate:发消息的工具类,也可以注入 JmsTemplate,JmsMessagingTemplate 是对 JmsTemplate 的封装。参数 destination 是发送到的队列,message 是待发送的消息。

11.3.2 创建消费者

生产者 AyMoodProducer 开发完成之后,我们来开发消费者 AyMoodConsumer,具体代码如下:

```
/**
 * 消费者
 * @author Ay
 * @date   2017/11/30
```

```
*/
@Component
public class AyMoodConsumer{

    @JmsListener(destination = "ay.queue")
    public void receiveQueue(String text) {
        System.out.println("用户发表说说【" + text + "】成功");
    }
}
```

- @JmsListener：使用JmsListener配置消费者监听的队列ay.queue，其中text是接收到的消息。

11.3.3 测试

生产者和消费者开发完成之后，我们在测试类 DemoApplicationTests 下开发测试方法 testActiveMQ ()，具体代码如下：

```
@Resource
private AyMoodProducer ayMoodProducer;

@Test
public void testActiveMQ() {

    Destination destination = new ActiveMQQueue("ay.queue");
    ayMoodProducer.sendMessage(destination, "hello,mq!!!");
}
```

测试方法开发完成之后，运行测试方法，可以在控制台看到打印的信息，如图 11-7 所示。同时可以在浏览器中访问 http://localhost:8161/admin 查看队列 ay.queue 的消费情况，具体如图 11-8 所示。

图 11-7　控制台打印信息　　　　　　　图 11-8　ay.queue 消费情况

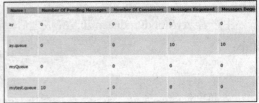

生产者和消费者开发完成之后，现在我们把用户发说说改成异步消费模式。首先在 AyMoodService 类下添加异步保存接口 asynSave()，具体代码如下：

```java
/**
 * 描述：微信说说服务层
 * @author Ay
 * @date   2017/12/2
 */
public interface AyMoodService {
    AyMood save(AyMood ayMood);
    String asynSave(AyMood ayMood);
}
```

然后在类 AyMoodServiceImpl 下实现 asynSave 方法，asynSave 方法并不保存说说记录，而是调用 AyMoodProducer 类的 sendMessage 推送消息，完整代码如下：

```java
/**
 * 描述：微信说说服务层
 * @author Ay
 * @date   2017/12/2
 */
@Service
public class AyMoodServiceImpl implements AyMoodService{

    @Resource
    private AyMoodRepository ayMoodRepository;
    @Override
    public AyMood save(AyMood ayMood) {
        return ayMoodRepository.save(ayMood);
    }
    //队列
    private static Destination destination = new ActiveMQQueue("ay.queue.asyn.save");

    @Resource
    private AyMoodProducer ayMoodProducer;
    @Override
    public String asynSave(AyMood ayMood){
        //往队列 ay.queue.asyn.save 推送消息，消息内容为说说实体
```

```
            ayMoodProducer.sendMessage(destination, ayMood);
            return "success";
        }
    }
```

其次，在 AyMoodProducer 生产者类下添加 sendMessage(Destination destination, final AyMood ayMood)方法，消息内容是 ayMood 实体对象。AyMoodProducer 生产者的完整代码如下：

```
/**
 * 生产者
 * @author Ay
 * @date 2017/11/30
 */
@Service
public class AyMoodProducer {

    @Resource
    private JmsMessagingTemplate jmsMessagingTemplate;

    public void sendMessage(Destination destination, final String message) {
        jmsMessagingTemplate.convertAndSend(destination, message);
    }

    public void sendMessage(Destination destination, final AyMood ayMood) {
        jmsMessagingTemplate.convertAndSend(destination, ayMood);
    }
}
```

最后，修改 AyMoodConsumer 消费者，在 receiveQueue 方法中保持说说记录，完整代码如下：

```
/**
 * 消费者
 * @author Ay
 * @date   2017/11/30
 */
@Component
public class AyMoodConsumer {

    @JmsListener(destination = "ay.queue")
```

```java
public void receiveQueue(String text) {
    System.out.println("用户发表说说【" + text + "】成功");
}

@Resource
private AyMoodService ayMoodService;

@JmsListener(destination = "ay.queue.asyn.save")
public void receiveQueue(AyMood ayMood){
    ayMoodService.save(ayMood);
}
}
```

用户发表说说，异步保存所有代码开发完成之后，我们在测试类 DemoApplicationTests 下添加 testActiveMQAsynSave 测试方法，具体代码如下：

```java
@Test
public void testActiveMQAsynSave() {
    AyMood ayMood = new AyMood();
    ayMood.setId("2");
    ayMood.setUserId("2");
    ayMood.setPraiseNum(0);
    ayMood.setContent("这是我的第一条微信说说！！！");
    ayMood.setPublishTime(new Date());
    String msg = ayMoodService.asynSave(ayMood);
    System.out.println("异步发表说说 :" + msg);
}
```

运行测试方法 testActiveMQAsynSave 成功之后，可以在数据库表 ay_mood 查询到用户 id 为 2 发表的记录，如图 11-9 所示。

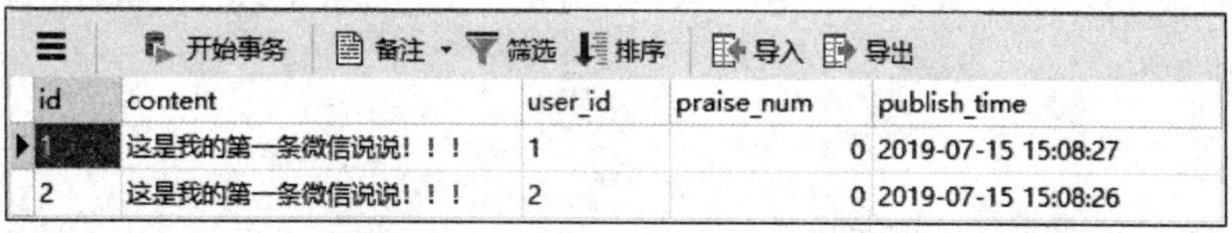

id	content	user_id	praise_num	publish_time
1	这是我的第一条微信说说！！！	1	0	2019-07-15 15:08:27
2	这是我的第一条微信说说！！！	2	0	2019-07-15 15:08:26

图 11-9 ay_mood 表记录

11.4　Spring Boot 异步调用

11.4.1　异步调用概述

异步调用是相对于同步调用而言的，同步调用是指程序按预定顺序一步步执行，每一步必须等到上一步执行完成之后才能执行，而异步调用则无须等待上一步程序执行完成即可执行。在日常开发的项目中，当访问的接口较慢或者做耗时任务时，不想程序一直卡在耗时任务上，想让程序能够并行执行，除了可以使用多线程来并行地处理任务，也可以使用 Spring Boot 提供的异步处理方式@Async 来处理。在 Spring Boot 框架中，只要添加@Async 注解就能将普通的同步任务改为异步调用任务。

11.4.2　@Async 使用

使用@Async 注解之前，需要在入口类添加注解@EnableAsync 开启异步调用，具体代码如下：

```
@SpringBootApplication
@ServletComponentScan
@ImportResource(locations={"classpath:spring-mvc.xml"})
@EnableAsync
public class MySpringBootApplication {

    public static void main(String[] args) {
        SpringApplication.run(MySpringBootApplication.class, args);
    }
}
```

然后，修改 AyUserServiceImpl 类的 findAll 方法，使它能够记录方法执行花费的时间，具体代码如下：

```
@Override
public List<AyUser> findAll() {
    try{
        System.out.println("开始做任务");
        long start = System.currentTimeMillis();
```

```
            List<AyUser> ayUserList = ayUserRepository.findAll();
            long end = System.currentTimeMillis();
            System.out.println("完成任务, 耗时: " + (end - start) + "毫秒");
            return ayUserList;
        }catch (Exception e){
            logger.error("method [findAll] error",e);
            return Collections.EMPTY_LIST;
        }
    }
}
```

11.4.3 测试

AyUserServiceImpl 类的方法 findAll()开发完成之后,在 DemoApplicationTests 测试类下开发测试方法 testAsync(),该方法调用 3 次 findAll(),并记录总共消耗的时间,由于现在是同步调用,所以代码按照顺序一步一步执行。testAsync 方法的具体代码如下:

```
@Test
public void testAsync(){
    long startTime = System.currentTimeMillis();
    System.out.println("第一次查询所有用户! ");
    List<AyUser> ayUserList = ayUserService.findAll();
    System.out.println("第二次查询所有用户! ");
    List<AyUser> ayUserList2 = ayUserService.findAll();
    System.out.println("第三次查询所有用户! ");
    List<AyUser> ayUserList3 = ayUserService.findAll();
    long endTime = System.currentTimeMillis();
    System.out.println("总共消耗: " + (endTime - startTime) + "毫秒");
}
```

测试方法 testAsync()开发完成之后,运行它,运行成功之后,可以在控制台看到如下的打印信息。

```
第一次查询所有用户!
开始做任务
完成任务, 耗时: 371 毫秒
第二次查询所有用户!
开始做任务
完成任务, 耗时: 34 毫秒
```

第三次查询所有用户！
开始做任务
完成任务，耗时：32 毫秒
总共消耗：438 毫秒

从打印结果可以看出，调用 3 次 findAll()，总共消耗 438 毫秒。现在在 AyUserService 接口中添加异步查询方法 findAsynAll()，并在 AyUserServiceImpl 类实现该方法，具体代码如下：

```java
/**
 * 描述：用户服务层接口
 * @author 阿毅
 * @date   2017/10/14
 */
public interface AyUserService {

    //省略大量代码

    List<AyUser> findAll();
    //异步查询
    Future<List<AyUser>> findAsynAll();
}
```

在 AyUserServiceImpl 类中实现 findAsynAll()方法，并在方法上添加异步调用注解@Async，具体代码如下：

```java
@Override
@Async
public Future<List<AyUser>> findAsynAll() {
    try{
        System.out.println("开始做任务");
        long start = System.currentTimeMillis();
        List<AyUser> ayUserList = ayUserRepository.findAll();
        long end = System.currentTimeMillis();
        System.out.println("完成任务，耗时：" + (end - start) + "毫秒");
        return new AsyncResult<List<AyUser>>(ayUserList) ;
    }catch (Exception e){
        logger.error("method [findAll] error",e);
        return new AsyncResult<List<AyUser>>(null);
    }
}
```

findAsynAll()方法开发完成之后,在 DemoApplicationTests 测试类下开发测试方法 testAsync2(),具体代码如下:

```java
@Test
public void testAsync2()throws Exception{
    long startTime = System.currentTimeMillis();
    System.out.println("第一次查询所有用户!");
    Future<List<AyUser>> ayUserList = ayUserService.findAsynAll();
    System.out.println("第二次查询所有用户!");
    Future<List<AyUser>> ayUserList2 = ayUserService.findAsynAll();
    System.out.println("第三次查询所有用户!");
    Future<List<AyUser>> ayUserList3 = ayUserService.findAsynAll();
    while (true){
        if(ayUserList.isDone() && ayUserList2.isDone() && ayUserList3.isDone()){
            break;
        }else {
            Thread.sleep(10);
        }
    }
    long endTime = System.currentTimeMillis();
    System.out.println("总共消耗: " + (endTime - startTime) + "毫秒");
}
```

测试方法 testAsync2()开发完成之后,运行它,运行成功之后,可以在控制台看到如下的打印信息。

```
第一次查询所有用户!
第二次查询所有用户!
第三次查询所有用户!
开始做任务
开始做任务
开始做任务
完成任务,耗时: 316 毫秒
完成任务,耗时: 316 毫秒
完成任务,耗时: 315 毫秒
总共消耗: 334 毫秒
```

从上面的打印结果可以看出,testAsync2 方法执行速度比 testAsync 方法快 104 毫秒(即 438 －334)。由此说明异步调用的速度比同步调用快。

第 12 章
全局异常处理与 Retry 重试

本章主要介绍 Spring Boot 全局异常使用、自定义错误页面、全局异常类开发、Retry 重试机制的介绍与使用等内容。

12.1 全局异常介绍

由于 Web 应用请求处理过程中发生错误是很常见的情况，所以 Spring Boot 为我们提供了一个默认的映射：/error，当处理中抛出异常之后，会转到该请求中处理，并且该请求有一个全局的错误页面用来展示异常内容。比如现在我们启动 spring-boot-book-v2 项目（启动项目之前，记得启动 Redis 服务和 ActiveMQ 服务），项目启动完成之后，在浏览器中随便输入一个访问地址，比如 http://localhost:8080/ayUser/testdddd，由于地址不存在，Spring Boot 会跳转到错误界面，如图 12-1 所示。

Whitelabel Error Page

This application has no explicit mapping for /error, so you are seeing this as a fallback.

Sat Dec 02 22:41:15 CST 2017
There was an unexpected error (type=Not Found, status=404).
No message available

图 12-1 Error 错误界面

虽然 Spring Boot 提供了默认的 error 错误页面映射，但是在实际应用中，图 12-1 所示的错误页面对用户来说并不友好，通常需要我们自己来实现异常提示。

12.2 Spring Boot 全局异常使用

12.2.1 自定义错误页面

我们知道，Spring Boot 的错误提示页面对用户体验并不好，这一节我们来实现自己的错误提示页面。首先，在 spring-boot-book-v2 项目/src/main/resources/static 目录下新建自定义错误页面 404.html，具体代码如下：

```html
<!DOCTYPE html>
<html lang="en">
<head>
    <meta charset="UTF-8">
    <title>Title</title>
</head>
<body>
<div class="text" style=" text-align:center;">
    主人，我累了，让我休息一会!!!
</div>
</body>
</html>
```

错误页面内容很简单，就是当访问路径不存在时，在页面中间显示一句话："主人，我累了，让我休息一会!!!"。当然，在真正的项目中，该错误页面样式会更加高大上。404 错误页面开发完成之后，在 spring-boot-book-v2 项目目录/src/main/java/com.example.demo 下新建包 error，并在 error 包下新建 ErrorPageConfig 配置类，具体代码如下：

```java
/**
 * 描述：自定义错误页面
 * @author Ay
 * @date 2017/12/02
 */
@Configuration
public class ErrorPageConfig {
```

```
    @Bean
    public WebServerFactoryCustomizer<ConfigurableWebServerFactory>
        webServerFactoryCustomizer(){
        return (container -> {
            ErrorPage error404Page = new ErrorPage(HttpStatus.NOT_FOUND, "/404.html");
            container.addErrorPages(error401Page, error404Page, error500Page);
        });
    }
```

- WebServerFactoryCustomizer：Spring Boot的自动配置有一个特性就是能够通过代码来修改配置，这样可以很方便地修改配置，而我们只需要实现Spring Boot定义的接口即可实现该功能。这里，需要注册一个实现了WebServerFactoryCustomizer的Bean，在ErrorPageConfig类中，使用匿名类来实现WebServerFactoryCustomizer接口，同时实现该接口唯一的方法customize，并自定义401、404、500等错误页面。

12.2.2 测试

404.html 错误页面和 ErrorPageConfig 类开发完成之后，重启动 spring-boot-book-v2 项目（项目启动之前，请记得启动 Redis 缓存服务和 ActiveMQ 服务，否则项目会报错，之后不再提示），在浏览器访问输入链接 http://localhost:8080/ayUser/testdddd，由于该链接不存在，就会出现如图 12-2 所示的自定义错误页面。

图 12-2　自定义 404 错误界面

12.2.3 全局异常类开发

在项目中，我们会遇到各种各样的业务异常，业务异常是指正常的业务处理时，由于某些业务的特殊要求而导致处理不能继续所抛出的异常。我们希望这些业务异常能够统一被处理，而使用 Spring Boot 进行全局异常处理很方便。首先，在目录 src/main/java/com.example.demo.error 下统一封装自定义业务异常类 BusinessException，该类继承 RuntimeException 异常类，并提供带有异常信息的构造方法，具体代码如下：

```
/**
* 描述：业务异常
* @author Ay
* @date   2017/12/3
*/
public class BusinessException extends RuntimeException{

    public BusinessException(){}

    public BusinessException(String message) {
        super(message);
    }
}
```

然后，在目录/src/main/java/com.example.demo.error 下新建错误信息类 ErrorInfo，该类用于封装错误信息，包括错误码，具体代码如下：

```
/**
* 描述：错误信息类
*@author Ay
* @date 2017/12/3
*/
public class ErrorInfo<T> {

    public static final Integer SUCCESS = 200;
    public static final Integer ERROR = 100;
    //错误信息
    private Integer code;
    //错误码
    private String message;
    private String url;
    private T data;
    //省略 set、get 方法
}
```

其次，在目录 /src/main/java/com.example.demo.error 下新建统一异常处理类 GlobalDefaultExceptionHandler，具体代码如下：

```
/**
 * 描述：统一业务异常处理类
 * @author Ay
 * @date   2017/12/3
 */
@ControllerAdvice(basePackages={"com.example.demo",})
public class GlobalDefaultExceptionHandler {

    @ExceptionHandler({BusinessException.class})
    //如果返回的为json数据或其他对象，添加该注解
    @ResponseBody
    public ErrorInfo defaultErrorHandler(HttpServletRequest req, Exception e) throws Exception {
        ErrorInfo errorInfo = new ErrorInfo();
        errorInfo.setMessage(e.getMessage());
        errorInfo.setUrl(req.getRequestURI());
        errorInfo.setCode(ErrorInfo.SUCCESS);
        return errorInfo;
    }

}
```

- @ControllerAdvice：定义统一的异常处理类，basePackages属性用于定义扫描哪些包，默认可不设置。
- @ExceptionHandler：用来定义函数针对的异常类型，可以传入多个需要捕获的异常类。
- @ResponseBody：如果返回的为json数据或其他对象，添加该注解。

最后，在AyUserController类下添加控制层方法findAll，并在方法里抛出BusinessException，该异常会被全局异常类捕获到，具体代码如下：

```
/**
 * 描述：用户控制层
 * @author Ay
 * @date   2017/10/28
 */
@Controller
@RequestMapping("/ayUser")
public class AyUserController {
```

```
@Resource
private AyUserService ayUserService;

@RequestMapping("/findAll")
public String findAll(Model model) {
    List<AyUser> ayUser = ayUserService.findAll();
    model.addAttribute("users",ayUser);
    //
    throw new BusinessException("业务异常");
}

//省略大量代码

}
```

12.2.4 测试

代码开发完成之后，重启动 spring-boot-book-v2 成功之后，在浏览器中输入访问地址：http://localhost:8080/ayUser/findAll，在浏览器中可以看到后端返回的 json 信息，具体信息如下：

{"code":200,"message":"业务异常","url":"/ayUser/findAll","data":null}

12.3 Retry 重试机制

12.3.1 Retry 重试概述

当我们调用一个接口时，由于网络等原因可能会造成第一次失败，再去尝试就成功了，这就是重试机制。重试的解决方案有很多，比如利用 try-catch-redo 简单重试模式，通过判断返回结果或监听异常来判断是否重试，具体请看如下简单的例子：

```
public void testRetry() throws Exception{
    boolean result = false;
    try{
        result = load();
        if(!result){
```

```
        load();//一次重试
    }
}catch (Exception e){
    load();//一次重试
}
}
```

try-catch-redo 重试模式还是有可能出现重试无效的现象，解决这个问题的方法是尝试增加重试次数 retrycount 和重试间隔周期 interval，以达到增加重试有效的可能性。因此我们可以利用 try-catch-redo-retry strategy 策略重试模式，具体伪代码如下所示：

```
public void testRetry2() throws Exception{
    boolean result = false;
    try{
        result = load();
        if(!result){
            //延迟3秒，重试3次
            reLoad(3000L,3);//延迟多次重试
        }
    }catch (Exception e){
        //延迟3秒，重试3次
        reLoad(3000L,3);//延迟多次重试
    }
}
```

但是这两种策略有一个共同的问题就是：正常逻辑和重试逻辑强耦合。基于这些问题，Spring-Retry 规范正常和重试逻辑，Spring-Retry 是一个开源工具包，该工具把重试操作模板定制化，可以设置重试策略和回退策略。同时重试执行实例保证线程安全。Spring-Retry 重试可以用 Java 代码方式实现，也可以用注解@Retryable 方式实现，这里 Spring-Retry 提倡以注解的方式对方法进行重试。

12.3.2　Retry 重试机制使用

使用 Spring 提供的重试策略之前，首先需要在 pom.xml 文件中引入所需的依赖，具体代码如下：

```xml
<dependency>
    <groupId>org.springframework.retry</groupId>
    <artifactId>spring-retry</artifactId>
</dependency>
<dependency>
    <groupId>org.aspectj</groupId>
    <artifactId>aspectjweaver</artifactId>
</dependency>
```

依赖添加完成之后，需要在入口类 MySpringBootApplication 添加注解@EnableRetry 开启 Retry 重试。完整代码如下：

```java
@SpringBootApplication
@ServletComponentScan
@ImportResource(locations={"classpath:spring-mvc.xml"})
@EnableAsync
//开启 Retry 重试
@EnableRetry
public class MySpringBootApplication {

    public static void main(String[] args) {
        SpringApplication.run(MySpringBootApplication.class, args);
    }
}
```

然后，在 AyUserService 类下添加新接口 findByNameAndPasswordRetry，具体代码如下：

```java
AyUser findByNameAndPasswordRetry(String name, String password);
```

接口 findByNameAndPasswordRetry 添加完成之后，在 AyUserServiceImpl 类下实现接口 findByNameAndPasswordRetry，并在方法中故意抛出业务异常 BusinessException，具体代码如下：

```java
@Override
@Retryable(value= {BusinessException.class},maxAttempts = 5,
backoff = @Backoff(delay = 5000,multiplier = 2))
public AyUser findByNameAndPasswordRetry(String name, String password) {
    System.out.println("[findByNameAndPasswordRetry] 方法失败重试了！");
    throw new BusinessException();
}
```

- @Retryable：value属性表示当出现哪些异常的时候触发重试，maxAttempts表示最大重试次数，默认为3，delay表示重试的延迟时间，multiplier表示上一次延时时间是这一次的倍数。

最后，在 AyUserController 类下添加控制层方法 findByNameAndPasswordRetry，在该方法中调用服务层 AyUserServiceImpl 的方法 findByNameAndPasswordRetry。具体代码如下：

```
@RequestMapping("/findByNameAndPasswordRetry")
public String findByNameAndPasswordRetry(Model model) {
    AyUser ayUser = ayUserService.findByNameAndPasswordRetry("阿毅","123456");
    model.addAttribute("users",ayUser);
    return "success";
}
```

12.3.3 测试

代码开发完成之后，重新启动 spring-boot-book-v2 项目，项目运行成功后，在浏览器中输入访问地址：http://localhost:8080/ayUser/findByNameAndPasswordRetry，由于方法 findByNameAndPasswordRetry 会抛出 BusinessException 异常，故 Retry 重试机制会检测到，进行第 2 次重试。重试成功，方法执行完成，重试失败，按照配置延迟 delay 时间，依次进行第 3 次、第 4 次、第 5 次重试，直到重试成功或者达到最大重试次数，重试策略终止。我们可以在 IDEA 的控制台多次看到如下的打印信息：

```
[findByNameAndPasswordRetry] 方法失败重试了！
```

第 13 章

集成 MongoDB 数据库

本章主要介绍如何安装和使用 MongoDB 数据库、NoSQL Manager for MongoDB 客户端安装与使用以及在 Spring Boot 中集成 MongoDB 数据库开发简单的功能等内容。

13.1 MongoDB 数据库介绍

13.1.1 MongoDB 的安装

MongoDB 是一个高性能、开源、无模式的文档型数据库,是当前 NoSQL 数据库中比较热门的一种,在企业中被广泛使用。其主要功能特性有:面向集合存储、易存储对象类型的数据、支持动态查询、文件存储格式为 BSON(一种 JSON 的扩展)、支持复制和故障恢复等。MongoDB 非常适合实时地插入、更新与查询,并具备网站实时数据存储所需的复制及高度伸缩性。由于性能很高,MongoDB 也适合作为信息基础设施的缓存层。在系统重启之后,由 MongoDB 搭建的持久化缓存层可以避免下层的数据源过载。由于高伸缩性,MongoDB 也非常适合由数十或数百台服务器组成的数据库。MongoDB 的路线图中已经包含对 MapReduce 引擎的内置支持,用于对象及 JSON 数据的存储方式,MongoDB 的 BSON 数据格式非常适合文档化格式的存储及查询。MongoDB 有很多优点,但缺点也是明显的,比如不能建立实体关系、没有事务管理机制等。

MongoDB 提供有 Windows、Linux、Mac OS X、Solaris 等操作系统的安装包，本书主要针对 Windows 操作系统进行讲解，具体安装步骤如下：

步骤01 在官方网站 https://www.mongodb.com/download-center#community 下根据自己的电脑操作系统的位数，下载对应的安装包，具体如图 13-1 所示。

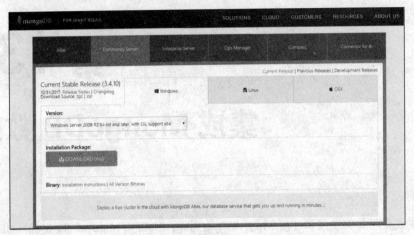

图 13-1　MongoDB 的官方网站

步骤02 双击下载的安装包，进行安装。这里笔者安装在 C 盘。

步骤03 找到安装目录的 bin 路径，将其配置在 Window 的环境变量 path 中，如 C:\Program Files\MongoDB\Server\3.4\bin。

步骤04 创建一个保存数据库的目录，如 C:\mongodb\data，然后打开一个命令窗口，输入命令：mongod --dbpath=c:\mongodb\data，启动 MongoDB 服务。

步骤05 安装成功之后，可以在命令行看到如图 13-2 所示的信息，代表 MongoDB 安装成功。

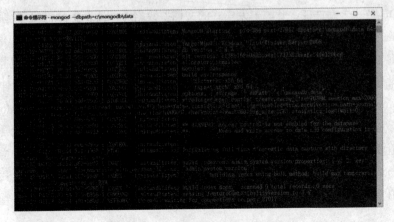

图 13-2　MongoDB 安装成功信息

13.1.2　NoSQL Manager for MongoDB 客户端的使用

连接 MongoDB 数据库的方式很多，除了可以使用最原始的命令窗口，还可以使用功能强大的 NoSQL Manager for MongoDB 客户端来连接 MongoDB 数据库。NoSQL Manager for MongoDB 客户端安装包可以到官方网站 https://www.mongodbmanager.com/download 下载。下载完成后按照正常的程序一步一步安装即可。

NoSQL Manager for MongoDB 客户端安装完成之后，在 MongoDB 数据库已启动的情况下，我们打开 NoSQL Manager for MongoDB 客户端，界面如图 13-3 所示。

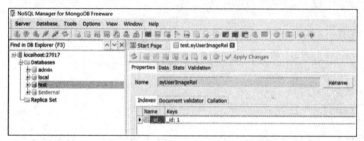

图 13-3　NoSQL Manager for MongoDB 界面

从图 13-3 中可以看出，MongoDB 数据库安装完成之后，默认创建了 3 个数据库，分别为 admin、local、test。单击 test 数据库，然后单击菜单栏的 shell 按钮，可以打开 shell 窗口，在 shell 窗口中，可以编写相关的 SQL 语句，具体如图 13-4 所示。

在 shell 菜单中输入 SQL 语句：show dbs，单击执行按钮，可以在 Result 结果界面看到 SQL 语句执行的结果，如图 13-5 所示。

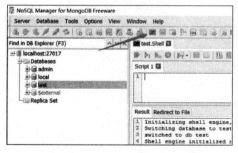

图 13-4　shell 菜单

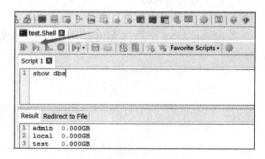

图 13-5　SQL 语句执行结果界面

还可以在 shell 窗口中编写如下的命令练习 MongoDB 的一些 SQL 语法，具体命令如下：

```
//显示所有的数据库
> show dbs;
```

```
//切换到 test 数据库
> use test;
//创建一个名为 user 的集合
> db.user.insert({"id":"1","name":"ay","age":"26"});
//查询 user 集合
> db.user.find();
//将集合 user 名称为 ay 的记录更新为名称为 al
> db.user.update({"id":"1"},{"id":"1","name":"al","age":"26"});
//查询 user 集合
> db.user.find();
//显示所有集合
> show collections;
//删除集合 user 名称为 al 的记录
> db.user.remove({"name":"al"});
//查询 user 集合
> db.user.find();
```

上面的命令只是一个简单的 SQL 语句练习,更多关于 MongoDB 的 SQL 语句练习,读者可到网上查阅相关资料学习。

13.2 集成 MongoDB

13.2.1 引入依赖

在 Spring Boot 中集成 MongoDB,首先需要在 pom.xml 文件中引入所需的依赖,具体代码如下:

```
<dependency>
    <groupId>org.springframework.boot</groupId>
    <artifactId>spring-boot-starter-data-mongodb</artifactId>
</dependency>
```

13.2.2 添加 MongoDB 配置

在 pom 文件引入 MongoDB 所需的依赖之后,需要在 application.properties 文件中添加如下的配置信息:

```
###MongoDB 配置
###host 地址
spring.data.mongodb.host=localhost
###默认数据库端口 27017
spring.data.mongodb.port=27017
###连接数据库名 test
spring.data.mongodb.database=test
```

13.2.3 连接 MongoDB

首先，在 my-spring-loot-v2 项目目录/src/main/java/com.example.demo.model 下新建用户附件类 AyUserAttachmentRel，具体代码如下：

```
/**
 * 描述：用户头像关联表
 * @author  Ay
 * @date    2017/12/4
 */
public class AyUserAttachmentRel {

    @Id
    private String id;
    private String userId;
    private String fileName;
    //省略 set、get 方法
}
```

用户附件类 AyUserAttachmentRel 开发完成之后，开发 AyUserAttachmentRelRepository 类，该类继承 MongoRepository 类，MongoRepository 类在 spring-data-mongodb 包下，类似于第 3 章的 Spring Data JPA。追溯 MongoRepository 源代码可以看出，MongoRepository 最顶级的父类就是 Repository 接口。AyUserAttachmentRelRepository 类的具体代码如下：

```
/**
 * 描述：用户附件 Repository
 * @author Ay
 * @date   2017/12/4
 */
```

```java
public interface AyUserAttachmentRelRepository
                extends MongoRepository<AyUserAttachmentRel,String> {

}
```

AyUserAttachmentRelRepository 类很简单，只是纯粹地继承 MongoRepository，就能继承 MongoRepository 为我们提供的增删改查等方法。AyUserAttachmentRelRepository 开发完成之后，我们开发服务层接口 AyUserAttachmentRelService 类，在 AyUserAttachmentRelService 类中声明 save 方法，用来简单保存数据，具体代码如下：

```java
/**
 * 描述：用户附件服务层
 * @author Ay
 * @date   2017/12/4
 */
public interface AyUserAttachmentRelService {

    AyUserAttachmentRel save(AyUserAttachmentRel ayUserAttachmentRel);

}
```

接口 AyUserAttachmentRelService 类开发完成之后，接下来开发其对应的实现类 AyUserAttachmentRelServiceImpl，在 AyUserAttachmentRelServiceImpl 中实现接口层方法 save，注入 AyUserAttachmentRelRepository 类，并调用 AyUserAttachmentRelRepository 的 save 方法将数据保存到 MongoDB 数据库中。具体代码如下：

```java
/**
 * 描述：用户附件实现层
 * @author Ay
 * @date   2017/12/4
 */
@Service
public class AyUserAttachmentRelServiceImpl implements AyUserAttachmentRelService {

    @Resource
    private AyUserAttachmentRelRepository ayUserAttachmentRelRepository;
```

```
    public AyUserAttachmentRel save(AyUserAttachmentRel ayUserAttachmentRel){
        return ayUserAttachmentRelRepository.save(ayUserAttachmentRel);
    }
}
```

13.2.4 测试

所有代码开发完成之后,在测试类 MySpringBootApplicationTests 下添加测试方法 testMongoDB,在方法中创建 AyUserAttachmentRel 实体,并调用 ayUserAttachmentRelService 类中的 save 方法,将数据存储到 MongoDB 中,具体代码如下:

```
@Resource
private AyUserAttachmentRelService ayUserAttachmentRelService;

@Test
public void testMongoDB(){
    AyUserAttachmentRel ayUserAttachmentRel = new AyUserAttachmentRel();
    ayUserAttachmentRel.setId("1");
    ayUserAttachmentRel.setUserId("1");
    ayUserAttachmentRel.setFileName("个人简历.doc");
    ayUserAttachmentRelService.save(ayUserAttachmentRel);
    System.out.println("保存成功");
}
```

运行测试方法 testMongoDB,运行前记得启动 MongoDB 数据库,测试方法运行成功之后,我们可以在 MongoDB 数据库查询到数据。具体查询的 SQL 语句如下:

```
> use test;
> db.ayUserImageRel.find();
```

查询结果:

```
{ "_id" : "1", "_class" : "com.example.demo.model.AyUserImageRel", "userId" : "1", "f
FileName" : "个人简历.doc" }
```

第 14 章

集成 Spring Security

本章主要介绍 Spring Security 的基础知识，Spring Boot 如何集成 Spring Security，利用 Spring Security 实现授权登录，以及利用 Spring Boot 实现数据库数据授权登录等内容。

14.1 Spring Security 概述

在 Web 应用开发中，安全是非常重要的。因为安全属于应用的非功能性需求，大部分企业更多地会把资源投入到应用开发中，所以应用安全很容易被忽略。安全应该在应用开发的初期就考虑进来，如果在应用开发的后期才考虑安全的问题，就可能陷入一个两难的境地：一方面，应用存在严重的安全漏洞，无法满足用户的要求，并可能造成用户的隐私数据被攻击者窃取；另一方面，应用的基本架构已经确定，要修复安全漏洞，可能需要对系统的架构做比较重大的调整，因而需要更多的开发时间，影响应用的发布进程。因此，从应用开发的第一天就应该把安全相关的因素考虑进来，并贯穿在整个应用的开发过程中。

市场上开源的安全框架很多，比如 Apache Shiro 安全框架、Spring Security 安全框架等。虽然 Spring Security 安全框架比 Apache Shiro 安全框架"重"，但是如果我们用心去了解 Spring Security 安全框架的话，会发现其实 Spring Security 安全框架是非常优秀的，所以本书选择在 Spring Boot 中集成 Spring Security 安全框架。

Spring Security 安全框架除了包含基本的认证和授权功能，还提供了加密解密、统一登录等一系列的支持。Spring Security 安全框架简单的实现原理如图 14-1 所示。

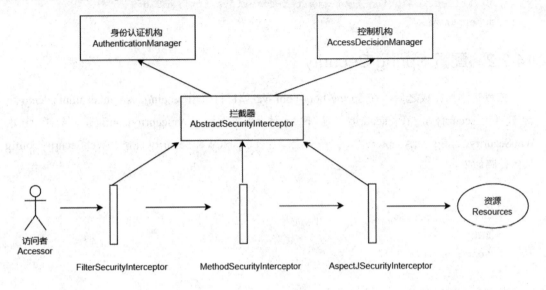

图 14-1　Spring Security 简单原理

在图 14-1 中，Accessor 是资源的访问者，在访问过程中需要经过一系列拦截器 Interceptor 的拦截，比如 FilterSecurityInterceptor、MethodSecurityInterceptor、AspectJSecurityInterceptor 等。这些拦截器是统一的抽象类 AbstractSecurityInterceptor 的具体实现。"控制机构" AccessDecisionManager 决定谁可以访问资源，而"身份认证机构" AuthenticationManager 就是定义那个"谁"，解决的是访问者身份认证的问题，只有确定注册类，才可以给予访问授权。"控制机构" AccessDecisionManager 和"身份认证机构" AuthenticationManager 负责制订规则，AbstractSecurityInterceptor 负责执行。

14.2　集成 Spring Security 的步骤

14.2.1　引入依赖

在 Spring Boot 中集成 Spring Security，首先需要在 pom.xml 文件中引入所需的依赖，具体代码如下：

```xml
<dependency>
    <groupId>org.springframework.boot</groupId>
    <artifactId>spring-boot-starter-security</artifactId>
</dependency>
```

14.2.2 配置 Spring Security

依赖包添加完成之后，在 spring-boot-book-v2 项目目录/src/main/java/com.example.demo 下新建包 security，在 security 包下新建配置类 WebSecurityConfig，该类继承 WebSecurityConfigurerAdapter 类，并在类上添加@EnableWebSecurity 注解。WebSecurityConfig 具体代码如下：

```java
/**
 * 描述: security 配置类
 * @author ay
 * @date   2017/12/10
 */
@Configuration
@EnableWebSecurity
public class WebSecurityConfig extends WebSecurityConfigurerAdapter {

    @Override
    protected void configure(HttpSecurity http) throws Exception {

        //路由策略和访问权限的简单配置
        http
                .formLogin()                                    //启用默认登录页面
                .failureUrl("/login?error")         //登录失败返回 URL:/login?error
                .defaultSuccessUrl("/ayUser/test ")
                                                                //登录成功跳转 URL，这里跳转到用户首页
                .permitAll();                                   //登录页面全部权限可访问
        super.configure(http);
    }

    /**
     * 配置内存用户
     */
    @Autowired
```

```
        public void configureGlobal(AuthenticationManagerBuilder auth) throws
Exception {
        auth
            .inMemoryAuthentication()
            .passwordEncoder(new MyPasswordEncoder())
            .withUser("阿毅").password("123456").roles("ADMIN")
            .and()
            .withUser("阿兰").password("123456").roles("USER");
    }
}
```

- @EnableWebSecurity：开启Security安全框架。
- configure方法：WebSecurityConfig继承WebSecurityConfigurerAdapter类需要重写configure方法，在方法中通过formLogin方法配置启用默认登录页面，通过failureUrl方法配置登录失败返回URL，通过defaultSuccessUrl配置登录成功跳转URL，这里调整到用户首页，通过permitAll方法设置登录页面全部权限可访问等。
- configureGlobal 方法：参数 AuthenticationManagerBuilder 类的方法 inMemoryAuthentication可添加内存中的用户，并可给用户指定角色权限。比如上面的代码给用户"阿毅"分配了ADMIN权限，而给用户"阿兰"分配了USER权限。

在 src/main/java/com.example.demo.security 目录下创建 MyPasswordEncoder 类，具体代码如下：

```
package com.example.demo.security;
import org.springframework.security.crypto.password.PasswordEncoder;

/**
 * 描述：
 * @author ay
 * @date 2019/07/07
 */
public class MyPasswordEncoder implements PasswordEncoder {
    @Override
    public String encode(CharSequence charSequence) {
        return charSequence.toString();
    }

    @Override
```

```
public boolean matches(CharSequence charSequence, String s) {
    return s.equals(charSequence.toString());
}
```
}

Spring Security 的 PasswordEncoder 接口用于执行密码的单向转换，以便安全地存储密码。简单来说，数据库存储的密码基本都是经过编码的，而决定如何编码以及判断未编码的字符序列和编码后的字符串是否匹配就是 PassswordEncoder 的责任。

14.2.3 测试

WebSecurityConfig 配置类开发完成，我们重新启动 spring-boot-book-v2 项目，项目启动成功后，在浏览器中输入访问链接：http://localhost:8080/ayUser/test，该访问请求会被 Security 框架拦截并跳转到默认的登录界面，具体如图 14-2 所示。在输入框中输入错误的用户名和密码：admin 和 123456。由于用户名和密码错误，Security 框架会跳转到之前在 WebSecurityConfig 类中配置的错误 URL:login?error 页面去，如图 14-3 所示。

图 14-2　Security 登录界面

图 14-3　登录错误界面

14.2.4 数据库用户授权登录

在 14.2.2 节中，用户登录系统的用户名、密码及角色都是写死在代码，显然不符合正常的逻辑。真正的项目都是通过查询数据库的用户名和密码进行用户认证和授权登录的。首先，需要在数据库中建立角色表 ay_role 和用户角色关联表 ay_user_role_rel，具体的建表语句如下：

```
-- 角色表
DROP TABLE IF EXISTS `ay_role`;
CREATE TABLE `ay_role` (
  `id` varchar(255) DEFAULT NULL,
```

```
  `name` varchar(255) DEFAULT NULL
);
-- 用户角色关联表
DROP TABLE IF EXISTS `ay_user_role_rel`;
CREATE TABLE `ay_user_role_rel` (
  `user_id` varchar(255) DEFAULT NULL,
  `role_id` varchar(255) DEFAULT NULL
);
```

角色表 ay_role 和用户角色关联表 ay_user_role_rel 建立完成之后，往表里插入数据：

```
-- 角色表 ay_role 数据
INSERT INTO `ay_role` VALUES ('1', 'ADMIN');
INSERT INTO `ay_role` VALUES ('2', 'USER');
-- 用户角色关联表 ay_user_role_rel
INSERT INTO `ay_user_role_rel` VALUES ('1', '1');
INSERT INTO `ay_user_role_rel` VALUES ('2', '2');
```

表 ay_role 和表 ay_user_role_rel 数据很简单，id 为 1 用户拥有 ADMIN 角色，id 为 2 用户拥有 USER 角色。数据插入完成之后，生成对应的实体类：AyRole 和 AyUserRoleRel，具体代码如下：

```
/**
 * 描述：用户角色实体
 * @author Ay
 * @date   2017/12/10
 */
@Entity
@Table(name = "ay_role")
public class AyRole {

    @Id
    private String id;
    private String name;
    //省略 set、get 方法
}

/**
 * 描述：用户角色关联
```

```
 * @author   Ay
 * @date     2017/12/10.
 */
@Entity
@Table(name = "ay_user_role_rel")
public class AyUserRoleRel {
    @Id
    private String userId;
    private String roleId;
    //省略 set、get 方法
}
```

实体类 AyRole 和 AyUserRoleRel 开发完成之后,我们一样利用 JPA 生成对应的 Repository 接口:AyRoleRepository 和 AyUserRoleRelRepository,具体代码如下:

```
/**
 * 描述:用户角色 Repository
 * @author Ay
 * @date   2017/12/10
 */
public interface AyRoleRepository extends JpaRepository<AyRole,String> {

}

/**
 * 描述:用户角色关联 Repository
 * @author Ay
 * @date   2017/10/14
 */
public interface AyUserRoleRelRepository extends JpaRepository<AyUserRoleRel,String> {

    List<AyUserRoleRel> findByUserId(@Param("userId")String userID);
}
```

AyUserRoleRelRepository 类提供 findByUserId 方法用来查询用户关联的角色数据。Repository 接口 AyRoleRepository 和 AyUserRoleRelRepository 开发完成之后,我们生成对应的 Service 接口:AyRoleService 和 AyUserRoleRelService。具体代码如下:

```
/**
 * 描述：用户角色Service
 * @author 阿毅
 * @date    2017/10/14
 */
public interface AyRoleService {
    AyRole find(String id);
}
```

AyRoleService 类提供 find 接口用来查询 AyRole 实体。

```
/**
 * 描述：用户角色关联Service
 * @author 阿毅
 * @date    2017/10/14
 */
public interface AyUserRoleRelService {
    List<AyUserRoleRel> findByUserId(String userId);
}
```

AyUserService 接口中添加 findByUserName 接口，用于根据用户名查询具体的用户。

```
AyUser findByUserName(String name);
```

在 AyUserServiceImpl 方法中简单实现 findByUserName 接口，具体代码如下：

```
@Override
public AyUser findByUserName(String name) {
    List<AyUser> ayUsers = findByName(name);
    if(ayUsers == null && ayUsers.size() <= 0){
        return null;
    }
    return ayUsers.get(0);
}
```

AyUserRoleRelService 类提供 findByUserId 接口用来查询用户关联角色实体。Service 接口 AyRoleService 和 AyUserRoleRelService 开发完成之后，我们开发对应的实现类：AyRoleServiceImpl 和 AyUserRoleRelServiceImpl，在实现类里实现 Service 接口，并注入对应的 Repository 接口，具体代码如下：

```java
/**
 * 描述: 用户角色 Service
 * @author Ay
 * @date    2017/12/2
 */
@Service
public class AyRoleServiceImpl implements AyRoleService {

    @Resource
    private AyRoleRepository ayRoleRepository;

    @Override
    public AyRole find(String id){
        return ayRoleRepository.findById(id).get();
    }
}


/**
 * 描述: 用户角色关联 Service
 * @author  Ay
 * @date    2017/12/10
 */
@Service
public class AyUserRoleServiceImpl implements AyUserRoleRelService{

    @Resource
    private AyUserRoleRelRepository ayUserRoleRelRepository;

    @Override
    public List<AyUserRoleRel> findByUserId(String userId) {
        return ayUserRoleRelRepository.findByUserId(userId);
    }
}
```

实现类 AyRoleServiceImpl 和 AyUserRoleRelServiceImpl 开发完成之后，我们需要开发 CustomUserService 类并实现 UserDetailsService 接口，UserDetailsService 接口是 Spring Security 框架提供的，CustomUserService 具体代码如下：

```java
/**
 * 描述：自定义用户服务类
 * @author Ay
 * @date    2017/12/10
 */
@Service
public class CustomUserService implements UserDetailsService{

    @Resource
    private AyUserService ayUserService;

    @Resource
    private AyUserRoleRelService ayUserRoleRelService;

    @Resource
    private AyRoleService ayRoleService;

    @Override
    public UserDetails loadUserByUsername(String name)
throws UsernameNotFoundException {
        AyUser ayUser = ayUserService.findByUserName(name);
        if(ayUser == null){
            throw new BusinessException("用户不存在");
        }
        //获取用户所有的关联角色
        List<AyUserRoleRel> ayRoleList
                = ayUserRoleRelService.findByUserId(ayUser.getId());
        List<GrantedAuthority> authorityList = new ArrayList<GrantedAuthority>();
        if(ayRoleList != null && ayRoleList.size() > 0){
            for(AyUserRoleRel rel:ayRoleList){
                //获取用户关联角色名称
                String roleName = ayRoleService.find(rel.getRoleId()).getName();
                authorityList.add(new SimpleGrantedAuthority(roleName));
            }
        }
        return new User(ayUser.getName(), ayUser.getPassword(), authorityList);
    }
}
```

在 CustomUserService 类中注入 AyUserService 服务类，并在 loadUserByUsername 方法中通过用户名查询用户。如果用户不存在，抛出业务异常 BusinessException，该异常类在第 12 章中已经开发完成，如果用户存在，继续根据用户 Id 查询用户关联的角色。最后在 loadUserByUsername 方法中返回 User 类，User 对象来自 org.springframework.security.core.userdetails.User，它实现了 UserDetails 接口，User 类的源代码如下：

```java
public class User implements UserDetails, CredentialsContainer {
    private static final long serialVersionUID = 420L;
    private String password;
    private final String username;

    public User(String username, String password, Collection<? extends GrantedAuthority> authorities) {
        this(username, password, true, true, true, true, authorities);
    }
    //省略大量代码
}
```

CustomUserService 类开发完成之后，需要在 WebSecurityConfig 类中注册，WebSecurityConfig 具体代码如下：

```java
/**
 * 描述: security 配置类
 * @author ay
 * @date  2017/12/10
 */
@Configuration
@EnableWebSecurity
public class WebSecurityConfig extends WebSecurityConfigurerAdapter {

    @Bean
    public CustomUserService customUserService(){
        return new CustomUserService();
    }

    @Override
    protected void configure(HttpSecurity http) throws Exception {
```

```java
        //路由策略和访问权限的简单配置
        http
            .formLogin()                              //启用默认登录页面
            .failureUrl("/login?error")               //登录失败返回URL:/login?error
            .defaultSuccessUrl("/ayUser/test")
                                                      //登录成功跳转URL，这里调整到用户首页
            .permitAll();                             //登录页面全部权限可访问
        super.configure(http);
    }
    /**
     * 配置内存用户
     */
    @Autowired
    public void configureGlobal(AuthenticationManagerBuilder auth) throws Exception {
        auth
            .userDetailsService(customUserService())
            .passwordEncoder(new MyPasswordEncoder());
//            .inMemoryAuthentication()
//            .withUser("阿毅").password("123456").roles("ADMIN")
//            .and()
//            .withUser("阿兰").password("123456").roles("USER");
    }
}
```

在 WebSecurityConfig 配置类中，通过注解@Bean 将 CustomUserService 装进 Spring 容器，并在 configureGlobal 方法中注册 CustomUserService 类。

14.2.5 测试

代码开发完成之后，重新启动 spring-boot-book-v2 项目，项目启动成功之后，在浏览器中输入访问链接：http://localhost:8080/login，在登录页面中输入用户名和密码：阿毅/123456，单击 Login 按钮登录，便可以登录成功。关闭浏览器或者清除浏览器的 Cookie 和缓存信息，重新访问链接：http://localhost:8080/login，在登录页面中输入用户名和密码：阿兰/123456，单击 Login 按钮登录，同样可以登录成功。

第 15 章

Spring Boot 应用监控

本章主要介绍如何通过 Spring Boot 监控和管理应用、自定义监控端点以及自定义 HealthIndicator 等内容。

15.1 应用监控介绍

Spring Boot 大部分模块都是用于开发业务功能或者连接外部资源。除此之外，Spring Boot 还提供了 spring-boot-starter-actuator 模块，该模块主要用于管理和监控应用，它是一个用于暴露自身信息的模块，可以有效地减少监控系统在采集应用指标时的开发量。

spring-boot-starter-actuator 模块提供了监控和管理端点以及一些常用的扩展和配置方式，具体如表 15-1 所示。

表 15-1 spring-boot-starter-actuator 模块的管理端点

路径（端点名）	描述	鉴权
/health	显示应用监控指标	false
/beans	查看 Bean 及其关系列表	true
/info	查看应用信息	false

(续表)

路径（端点名）	描述	鉴权
/trace	查看基本追踪信息	true
/env	查看所有环境变量	true
/env/{name}	查看具体变量值	true
/mappings	查看所有 URL 映射	true
/autoconfig	查看当前应用的所有自动配置	true
/configprops	查看应用所有配置属性	true
/shutdown	关闭应用（默认关闭）	true
/metrics	查看应用基本指标	true
/metrics/{name}	查看应用具体指标	true
/dump	打印线程栈	true

15.2 使用监控

15.2.1 引入依赖

在 Spring Boot 中使用监控，首先需要在 pom.xml 文件中引入所需的依赖 spring-boot-starter-actuator，具体代码如下：

```
<dependency>
    <groupId>org.springframework.boot</groupId>
    <artifactId>spring-boot-starter-actuator</artifactId>
    <version>1.5.10.RELEASE</version>
</dependency>
```

15.2.2 添加配置

在 pom.xml 文件引入 spring-boot-starter-actuator 依赖包之后，需要在 application.properties 文件中添加如下的配置信息：

```
### 应用监控配置
#指定访问这些监控方法的端口
management.server.port=8080
```

- management.server.port：用于指定访问这些监控方法的端口。

15.2.3 测试

spring-boot-starter-actuator 依赖和配置都添加成功之后,重新启动 spring-boot-book-v2 项目,项目启动成功之后,在浏览器中输入 http://localhost:8080/actuator,可以看到如图 15-1 所示的输出信息。

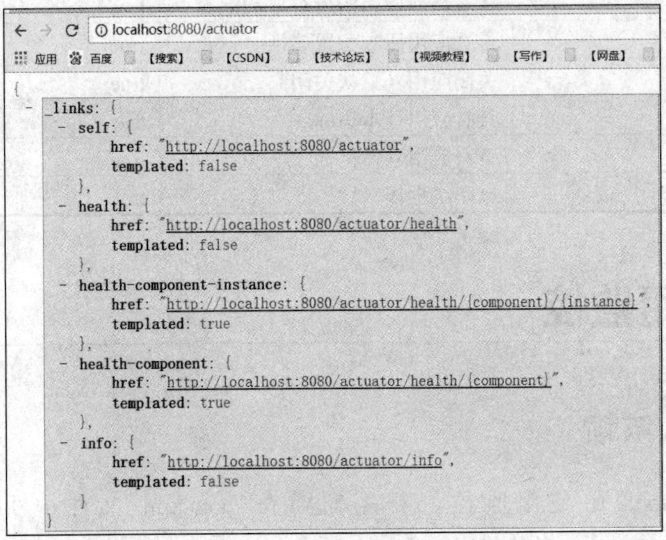

图 15-1 查看默认提供的 endpoint

从图 15-1 中可以看出,actuator 只暴露了 3 个简单的 endpoint,并且只有 /health 接口的内容还有点用,可以检查应用服务是否健康。当然,actuator 绝对不止这么点功能,只是出于安全考虑,其余的 endpoint 默认被禁用了。在浏览器中输入:http://localhost:8080/actuator/health,可以看到如图 15-2 所示的应用健康信息,"UP" 代表应用是健康状态。

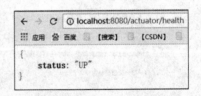

图 15-2 health 健康信息图

为了简单起见,我们来开启所有的接口。只需要在 application.properties 文件中加入一行配置即可:

```
### 开启所有的端点
management.endpoints.web.exposure.include=*
```

重新在浏览器中输入 http://localhost:8080/actuator,便可以看到所有的 endpoint,如图 15-3 所示。

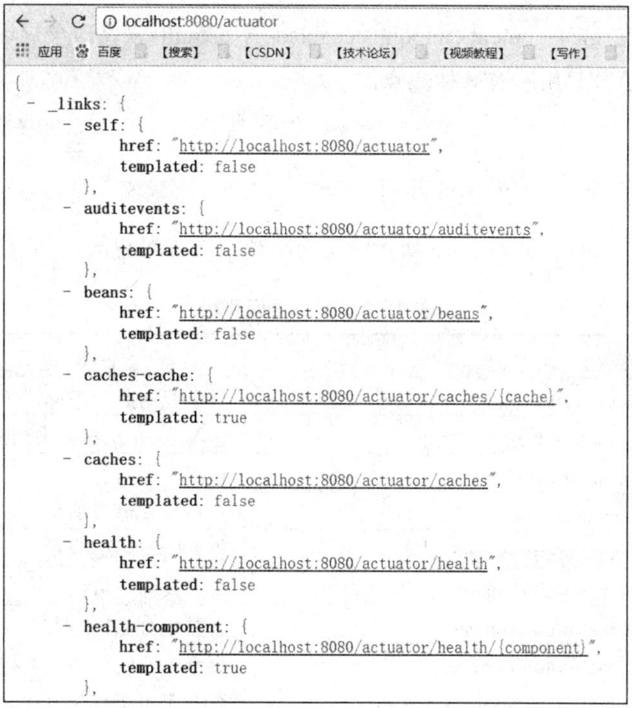

图 15-3 所有的 endpoint 信息

对于不带任何参数的读取操作，端点自动缓存对其响应。要配置端点缓存响应的时间，请使用 cache.time-live 属性，以下示例将 beans 端点缓存的生存时间设置为 10 秒：

```
management.endpoint.beans.cache.time-to-live=10s
```

默认情况下，端点通过使用端点的 ID 在/actuator 路径下的 HTTP 上公开，例如，beans 端点暴露在 /actuator/beans 下。如果要将端点映射到其他路径，则可以使用 management.endpoints.web.path-mapping 属性。另外，如果想更改基本路径，则可以使用 management.endpoints.web.base-path。以下示例将/actuator/health 重新映射到/healthcheck：

```
management.endpoints.web.base-path=/
management.endpoints.web.path-mapping.health=healthcheck
```

要配置单个端点的启用，请使用 management.endpoint.<id>.enabled 属性，以下示例启用了 shutdown 端点：

```
management.endpoint.shutdown.enabled=true
```

另外，可以通过 management.endpoints.enabled-by-default 来修改全局端口的默认配置，以下示例启用 info 端点并禁用所有其他端点：

```
management.endpoints.enabled-by-default=false
management.endpoint.info.enabled=true
```

其他端点测试，可以按照表 15-2 所示的访问路径依次访问测试。

表 15-2　端点访问路径

路径（端点名）	描　　述
http://localhost:8080/actuator/health	显示应用监控指标
http://localhost:8080/actuator/beans	查看 bean 及其关系列表
http://localhost:8080/actuator/info	查看应用信息
http://localhost:8080/actuator/trace	查看基本追踪信息
http://localhost:8080/actuator/env	查看所有环境变量
http://localhost:8080/actuator/env/{name}	查看具体变量值
http://localhost:8080/actuator/mappings	查看所有 url 映射
http://localhost:8080/actuator/autoconfig	查看当前应用的所有自动配置
http://localhost:8080/actuator/configprops	查看应用所有配置属性
http://localhost:8080/actuator/shutdown	关闭应用（默认关闭）
http://localhost:8080/actuator/metrics	查看应用基本指标
http://localhost:8080/actuator/metrics/{name}	查看应用具体指标
http://localhost:8080/actuator/dump	打印线程栈

在浏览器中可以把返回的数据格式化成 json 格式，这是因为在 Google 浏览器中安装了 JsonView 插件，具体安装步骤如下：

步骤 01　浏览器中输入链接：https://github.com/search?utf8=%E2%9C%93&q=jsonview，在弹出的页面中单击 gildas-lormeau/JSONView-for-Chrome，如图 15-4 所示。

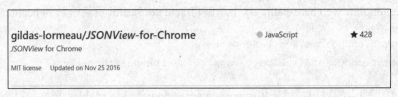

图 15-4　gildas-lormeau/JSONView-for-Chrome 界面

步骤 02　单击【Download Zip】，插件下载完成，解压缩到相应目录中（D:\Download\JSONView-for-Chrome-master）。

步骤03 在浏览器右上角单击【更多工具】→【扩展程序】→【加载已解压的扩展程序】。选择插件目录（D:\Download\JSONView-for-Chrome-master\WebContent）。

步骤04 安装完成后，重新启动浏览器（快捷键 Ctrl+R）。

15.3 自定义端点

15.3.1 自定义端点 EndPoint

spring-boot-starter-actuator 模块中已经提供了许多原生端点。根据端点的作用，我们可以把原生端点分为以下 3 大类。

- 应用配置类：获取应用程序中加载的应用配置、环境变量、自动化配置报告等与 Spring Boot 应用密切相关的配置类信息。
- 度量指标类：获取应用程序运行过程中用于监控的度量指标，比如内存信息、线程池信息、HTTP 请求统计等。
- 操作控制类：提供了对应用的关闭等操作类功能。

如果 spring-boot-starter-actuator 模块提供的这些原生端点无法满足需求，还可以自定义端点，自定义端点时，只要继承抽象类 AbstractEndpoint 即可。这里在 spring-boot-book-v2 目录 /src/main/java/com.example.demo 下新建 actuator 包，在 actuator 包下新建自定义端点类 AyUserEndPoint，AyUserEndPoint 主要用来监控数据库用户信息情况，比如用户总数量、被删除用户数量、活跃用户数量等。自定义端点 AyUserEndPoint 类的代码如下：

```
/**
 * 描述：自定义端点
 * @author  Ay
 * @date    2017/12/9
 */
@Component
@Endpoint(id="userEndPoints")
public class AyUserEndPoint{

    @Resource
    private AyUserService ayUserService;
```

```
    @ReadOperation
    public Map<String, Object> invoke() {
        Map<String, Object> map = new HashMap<String, Object>();
        //当前时间
        map.put("current_time",new Date());
        //用户总数量
        map.put("user_num",ayUserService.findUserTotalNum());
        return map;
    }
}
```

- @Endpoint(id="userEndPoints"): @Endpoint注解简化了创建用户自定义端点的过程，@Endpoint相当于@WebEndpoint和@JmxEndpoint的整合，Web和jmx方式都支持。
- @WebEndpoint: 只会生成Web的方式的端点监控。
- @JmxEndpoint: 只会生成Jmx的方式监控。

在 AyUserEndPoint 类中，我们注入 AyUserService 接口，并在 invoke 方法中调用 findUserTotalNum 方法，查询当前数据库总的用户数。所以需要在 AyUserService 接口中添加方法 findUserTotalNum，具体代码如下：

```
//查询用户数量
Long findUserTotalNum();
```

同时，在 AyUserServiceImpl 类中实现方法 findUserTotalNum，具体代码如下：

```
@Override
public Long findUserTotalNum() {
    return ayUserRepository.count();
}
```

AyUserService 类与 AyUserServiceImpl 类开发完成之后，就可以在 invoke 方法中使用。在 invoke 方法中定义 Map 集合，并向 Map 集合存放当前时间 current_time 和数据库用户总数 user_num。

15.3.2　测试

代码开发完成之后，重启启动 spring-boot-book-v2 项目，在浏览器中输入访问地址：http://localhost:8080/actuator/userEndPoints，便可以看到请求到数据，具体数据如下：

```
{"user_num":3,"current":1512817762910}
```

从返回数据中,可以看出当前数据库总共有 3 个用户,以及当前具体时间(毫秒)。

15.3.3 自定义 HealthIndicator

默认端点 Health 的信息是从 HealthIndicator 的 bean 中收集的,Spring 中内置了一些 HealthIndicator,如表 15-3 所示。

表 15-3 Spring 内置的 HealthIndicator

名称	描述
CassandraHealthIndicator	检测 Cassandra 数据库是否运行
DiskSpaceHealthIndicator	检测磁盘空间
DataSourceHealthIndicator	检测 DataSource 连接是否能获得
ElasticsearchHealthIndicator	检测 Elasticsearch 集群是否在运行
JmsHealthIndicator	检测 JMS 消息代理是否在运行
MailHealthIndicator	检测邮箱服务器是否在运行
MongoHealthIndicator	检测 Mongo 是否在运行
RabbitHealthIndicator	检测 RabbitMQ 是否在运行
RedisHealthIndicator	检测 Redis 是否在运行
SolrHealthIndicator	检测 Solr 是否在运行

启动项目 spring-boot-book-v2,在浏览器中输入访问链接:http://localhost:8080/actuator/health,可以看到返回的 Spring Boot 应用健康数据只有:

```
{
    "status":"UP"
}
```

如果想要查看详细的应用健康信息,需要添加以下配置:

```
management.endpoint.health.show-details=always
```

配置完成之后,再次访问 http://localhost:8080/actuator/health,获取的信息如下:

```
{
    status: "DOWN",
    details: {
        my: {
```

```
            status: "UP",
            details: {
                status: "UP",
                total: 10000,
                free: 5000
            }
        },
        diskSpace: {
            status: "UP",
            details: {
                total: 214748360704,
                free: 126022668288,
                threshold: 10485760
            }
        },
        mongo: {
            status: "UP",
            details: {
                version: "3.4.7"
            }
        },
        jms: {
            status: "UP",
            details: {
                provider: "ActiveMQ"
            }
        },
        db: {
            status: "UP",
            details: {
                database: "MySQL",
                hello: 1
            }
        },
        mail: {
            status: "DOWN",
            details: {
                location: "smtp.163.com:-1",
```

```
                error: "javax.mail.AuthenticationFailedException: 550 User has no
permission "
            }
        },
        redis: {
            status: "UP",
            details: {
                version: "3.0.503"
            }
        }
    }
}
```

从上面的信息中，可以方便地查看目前应用所依赖资源（Redis、MongoDB）的运行情况及其他信息。

 management.endpoint.health.show-details 的值除了 always 之外还有 when-authorized、never，默认值是 never。

如果想要自定义符合自己业务需求的检查健康，需要自定义 HealthIndicator 来获得更多应用健康的信息。在 spring-boot-book-v2 项目目录 /src/main/java/com.example.actuator 下新建 MyHealthIndicator 类，该类实现 HealthIndicator 接口并重写 health 方法，MyHealthIndicator 类具体代码如下：

```
/**
 * 描述：自定义健康监控类
 * @author Ay
 * @date   2017/12/9
 */
@Component
public class MyHealthIndicator implements HealthIndicator{

    @Override
    public Health health() {
        Long totalSpace = checkTocalSpace();
        Long free = checkFree();
        String status = checkStatus();
        checkFree();
```

```
            return new Health.Builder()
                    .up()
                    .withDetail("status",status)
                    .withDetail("total",totalSpace)
                    .withDetail("free",free)
                    .build();
        }
        private String checkStatus(){
            //结合真实项目，获取相关参数
            return "UP";
        }
        private Long checkTocalSpace(){
            //结合真实项目，获取相关参数
            return 10000L;
        }
        private Long checkFree(){
            //结合真实项目，获取相关参数
            return 5000L;
        }
    }
```

15.3.4　测试

代码开发完成之后，重新启动 spring-boot-book-v2 项目，项目启动成功之后，在浏览器中输入访问链接：http://localhost:8080/actuator/health，可以获得自定义健康类 MyHealthIndicator 返回的结果，具体结果信息如下：

```
my: {
    status: "UP",
    total: 10000,
    free: 5000
}
//省略其他健康数据
```

从上面返回的 json 结果信息可以看出，json 结果信息的 key：my，也就是英文 MyHealthIndicator 去掉 HealthIndicator。如果自定义健康类取名为 MyDefineHealthIndicator，则返回结果信息将会变成：

```
myDefine: {
status: "UP",
total: 10000,
free: 5000
}
//省略其他健康数据
```

一般情况下，不会直接实现 HealthIndicator 接口，而是继承 AbstractHealthIndicator 抽象类。因此，我们只需要重写 doHealthCheck 方法，并在这个方法中关注具体的健康检测的业务逻辑服务即可。

15.4　保护 Actuator 端点

Actuator 端点发布的信息很多都涉及敏感信息和高危操作。比如/shutdown 端点，它可以直接关闭应用程序，如果随便某个人都有权限访问该端点，那是非常危险的。因此，有必要控制 Actuator 端点的访问权限以避免 Actuator 端点被非法访问。想要保护 Actuator 端点，可以使用保护其他 URL 路径一样的方式，通过使用 Spring Security 来控制 URL 路径的授权访问。

在第 14 章中，我们已经在 Spring Boot 中集成了 Spring Security，并且开发了 WebSecurityConfig 配置类对用户登录进行授权访问，现在我们改造该类，具体代码如下：

```
/**
 * 描述：Security 配置类
 * @author ay
 * @date   2017/12/10
 */
@Configuration
@EnableWebSecurity
public class WebSecurityConfig extends WebSecurityConfigurerAdapter {

    //省略代码

    @Override
    protected void configure(HttpSecurity http) throws Exception {

        //路由策略和访问权限的简单配置
```

```
    http
        .authorizeRequests()
        //要求有管理员的权限
        .antMatchers("/shutdown").access("hasRole('ADMIN')")
        .antMatchers("/**").permitAll()
        .and()
        .formLogin()                       //启用默认登录页面
        .failureUrl("/login?error")        //登录失败返回URL:/login?error
        .defaultSuccessUrl("/ayUser/test")
                                           //登录成功跳转URL,这里跳转到用户首页
        .permitAll();                      //登录页面全部权限可访问
    super.configure(http);
    }
}
```

通过使用 antMatchers("/shutdown").access("hasRole('READER')")方法，对/shutdown 进行授权访问，/shutdown 端点现在仅允许拥有 ADMIN 权限的用户进行访问。

端点/shutdown 已经被保护起来了，假如现在想保护其他端点，例如/metrics、/health 等，只需要为 antMatchers()传入输入参数即可。具体代码如下：

```
.authorizeRequests()
//要求有管理员的权限
.antMatchers("/shutdown","/metrics","/health") .access("hasRole('READER')")
```

如果觉得每次添加一个端点的访问权限都得在 antMatchers()方法中修改很麻烦，可以在 application.properties 配置文件中配置端点访问的上下文，具体配置如下：

```
###配置端点访问的上下文路径
management.endpoints.web.base-path=/manage
```

此时，在为 Actuator 端点赋予 ADMIN 权限限制的时候就能借助这个上下文/manage：

```
//要求有管理员的权限
.antMatchers("/manage/**") .access("hasRole('READER')")
```

第 16 章

集成 Dubbo 和 Zookeeper

本章主要介绍如何安装并运行 Zookeeper，Spring Boot 集成 Dubbo，spring-boot-book-v2 项目的服务拆分和实践，正式版 API 如何发布，服务注册等内容。

16.1 Zookeeper 的介绍与安装

16.1.1 Zookeeper 概述

ZooKeeper 是一个开源的、分布式应用程序协调服务，它提供的功能包括命名服务、配置管理、集群管理、分布式锁等。

（1）命名服务。可以简单理解为电话簿。电话号码不好记，但是人名好记。要打谁的电话，直接查人名就好了。分布式环境下，经常需要对应用/服务进行统一命名，便于识别不同的服务。类似于域名与 IP 之间的对应关系，域名容易记住。ZooKeeper 通过名称来获取资源和服务的地址以及提供者等信息。

（2）配置管理。分布式系统都有大量的服务器，比如在搭建 Hadoop 的 HDFS 的时候，需要在一台 Master 主机器上配置好 HDFS 需要的各种配置文件，然后通过 scp 命令把这些配置文件拷贝到其他节点上，这样各个机器拿到的配置信息是一致的，才能成功运行 HDFS 服务。Zookeeper 提供了这样的一种服务，集中管理配置的方法，在这个集中的地方修改了配置，所有对这个配置感兴趣的都可以获得变更。这样就省去了手动复制配置，还保证了可靠和一致性。

（3）集群管理。所谓集群管理包含两点：是否有机器退出和加入、选举 Master。在分布式集群中，经常由于各种原因（比如硬件故障、软件故障、网络问题等），有些新的节点会加入进来，也有老的节点会退出集群。这个时候，集群中有些机器（比如 Master 节点）需要感知到这种变化，然后根据这种变化做出对应的决策。Zookeeper 集群管理可以感知变化并给出对应的策略。

（4）分布式锁。Zookeeper 的一致性文件系统，使得锁的问题变得容易了。锁服务可以分为两类，一类是保持独占，另一类是控制时序。单机程序的各个进程需要对互斥资源进行访问时需要加锁，分布式程序分布在各个主机上的进程对互斥资源进行访问时也需要加锁。很多分布式系统有多个可服务的窗口，但是在某个时刻只让一个服务去干活，当这台服务出问题的时候锁释放，立即 fail over 到另外的服务。在很多分布式系统中都是这么做的，这种设计有一个更好听的名字叫 Leader Election（leader 选举）。举个通俗点的例子，比如银行取钱，有多个窗口，但是，只能有一个窗口对你服务。如果正在对你服务的窗口的柜员突然有急事走了，怎么办呢？找大堂经理（Zookeeper），大堂经理会为你指定另外的窗口继续为你服务。

Zookeeper 一个最常用的使用场景是用于担任服务生产者和服务消费者的注册中心，这也是接下来的章节中会使用到的。服务生产者将自己提供的服务注册到 Zookeeper 中心，服务消费者在进行服务调用的时候先到 Zookeeper 中查找服务，获取到服务生产者的详细信息之后，再去调用服务生产者的内容与数据。具体如图 16-1 所示。

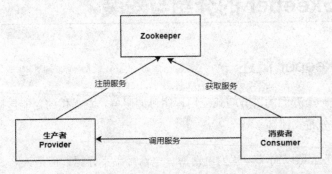

图 16-1　Zookeeper 服务注册简单原理

限于篇幅，这里只对 Zookeeper 进行简单的功能介绍。

16.1.2　Zookeeper 的安装与启动

Zookeeper 的安装与启动非常简单，具体步骤如下：

第 16 章 集成 Dubbo 和 Zookeeper

步骤 01 在官方网站 https://archive.apache.org/dist/zookeeper/ 下载对应的安装包，这里选用 zookeeper-3.3.6 版本。

步骤 02 将下载好的安装包解压到 D 盘目录下。在解压后的目录中找到配置文件 zoo_sample.cfg（存放在 D:\zookeeper-3.3.6\zookeeper-3.3.6\conf\目录下），该配置文件是 Zookeeper 为提供的最简单的配置文件，我们复制一份并重新命名为 zoo.cfg，Zookeeper 启动时会默认在 conf 目录下读取 zoo.cfg 配置文件，具体如图 16-2 所示。

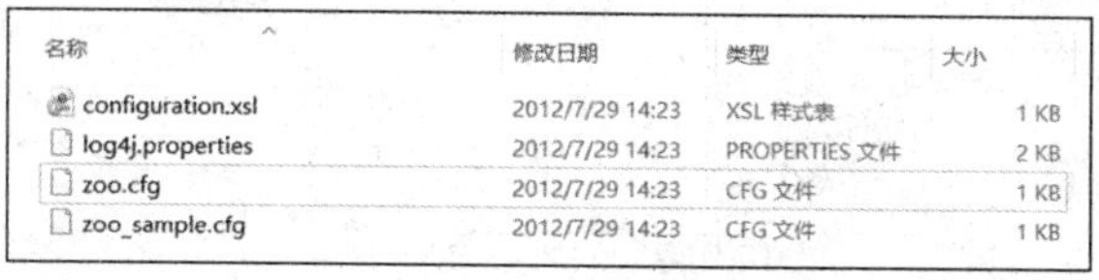

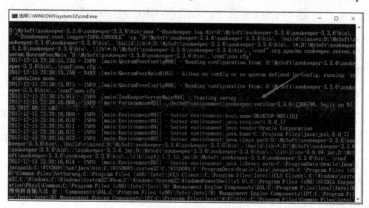

图 16-2 生成 zoo.cfg 配置文件

步骤 03 zoo.cfg 配置文件生成之后，在目录 D:\zookeeper-3.3.6\zookeeper-3.3.6\bin 下双击 zkServer.cmd 文件启动 zookeeper（如果是 Linux 环境，执行 zkServer.sh 文件）。在命令行窗口中看到如图 16-3 所示 Zookeeper 的启动信息，代表 Zookeeper 安装成功。

图 16-3 Zookeeper 安装成功信息

16.2 Spring Boot 集成 Dubbo

16.2.1 Dubbo 概述

Dubbo 是阿里巴巴 B2B 平台技术部开发的一款 Java 服务平台框架以及 SOA 治理方案。其功能主要包括：高性能 NIO 通信及多协议集成、服务动态寻址与路由、软负载均衡与容错、依赖分析与降级等。Dubbo 简单的底层框架如图 16-4 所示。

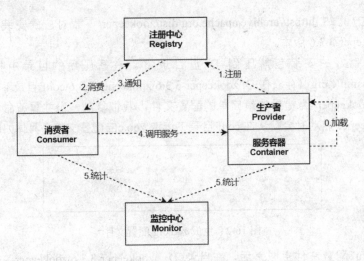

图 16-4　Dubbo 底层框架原理

Registry 是服务注册与发现的注册中心，Provider 是暴露服务的服务提供方，Consumer 是调用远程服务的服务消费方，Monitor 是统计服务的调用次数和调用时间的监控中心，Container 是服务运行容器。Dubbo 简单的调用关系如下：

（1）服务容器 Container 负责启动、加载、运行服务提供者 Provider。

（2）服务提供者 Provider 在启动时，向注册中心 Registry 注册自己提供的服务。

（3）服务消费者 Consumer 在启动时，向注册中心 Registry 订阅自己所需的服务。

（4）注册中心 Registry 返回服务提供者地址列表给消费者 Provider，如果有变更，注册中心 Registry 将基于长连接推送变更数据给消费者 Consumer。

（5）服务消费者 Consumer 从提供者地址列表中，基于软负载均衡算法，选一台提供者进行调用，如果调用失败，再选另一台调用。

（6）服务消费者 Consumer 和提供者 Provider 在内存中累计调用次数和调用时间，定时每分钟发送一次统计数据到监控中心 Monitor。

Dubbo 将注册 Registry 中心进行抽象，使得它可以外接不同的存储媒介给注册中心 Registry 提供服务，可以作为存储媒介的有 ZooKeeper、Memcached、Redis 等。引入 ZooKeeper 作为存储媒介，也就把 ZooKeeper 的特性引进来。首先是负载均衡，单注册中心的承载能力是有限的，在流量达到一定程度的时候就需要分流，负载均衡就是为了分流而存在的。一个 ZooKeeper 群配合相应的 Web 应用就可以很容易达到负载均衡；然后是资源同步，单单有负载均衡还不够，节点之间的数据和资源需要同步，ZooKeeper 集群天然具备这样的功能；最后是命名服务，将树状结构用于维护全局的服务地址列表，服务提供者在启动的时候，向 ZooKeeper 上的指定节点目录下写入自己的 URL 地址,这个操作就完成了服务的发布。ZooKeeper 的其他特性还有 Mast 选举、分布式锁等。ZooKeeper 的知识在 16.1 节中已经有简单的介绍,这里就不再重复赘述。

16.2.2 服务与接口拆分思路

截至 16 章,spring-boot-book -v2 项目已经集成了很多技术,也定义了很多的接口。但是对于真实的项目来说,特别是对于互联网公司项目来说,spring-boot-book-v2 这个大的服务承载的内容太多,诸多服务接口(比如 AyUserService、AyMoodService、AyRoleService 等)糅合在一块对外提供服务,已经违背了微服务理念。因此,我们有必要对 spring-boot-book-v2 项目进行服务拆分,使它被拆分成一个个小的服务,我们可以按照业务或者功能维度对服务进行拆分。具体如图 16-5 所示。

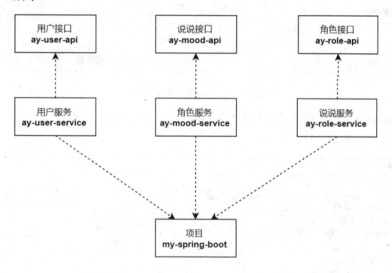

图 16-5 spring-boot-book-v2 服务拆分

在图 16-5 中,spring-boot-book-v2 项目被拆分为用户服务、角色服务、说说服务等。spring-boot-book-v2 项目依赖于这些底层的服务为其提供相应的功能,而用户服务、角色服务和说说服务是面向接口 API 编程,符合基本的编程原则。通过服务的拆分和面向接口编程,对于项目扩展和团队分工都有莫大的好处。

16.2.3 服务与接口拆分实践

我们已经清楚了服务拆分的原因和好处,本节在项目中实践它。首先,在 spring-boot-book-v2 项目下添加 ay-user-api、ay-mood-api、ay-role-api 接口模块。具体步骤如下:

步骤 01 选择 spring-boot-book-v2【右键】→【New】→【Module】，在弹出的窗口中选择【Spring Initializr】，然后单击【Next】按钮，具体如图 16-6 所示。

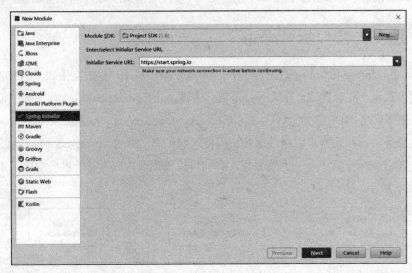

图 16-6　New Module 窗口

步骤 02 在 Name 输入框中输入模块的名称 ay-user-api，其他选项按照默认即可。然后单击【Next】按钮，具体如图 16-7 所示。

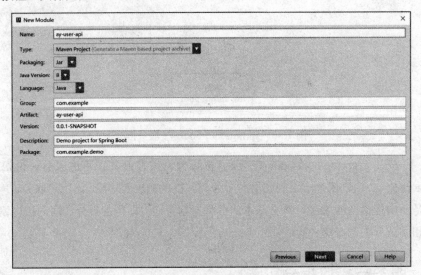

图 16-7　输入模块名称

步骤 03 一路单击【Next】按钮，在 Module name 输入框中输入模块的名称 ay-user-api，然后单击【Finish】按钮，具体如图 16-8 所示。

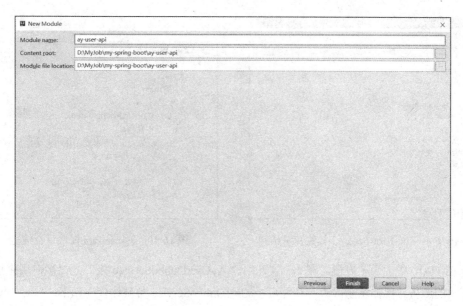

图 16-8　输入模块名称

按照上面的步骤建立好 ay-user-api 模块。再按照相同的步骤，依次在 spring-boot-book-v2 项目下创建接口模块：ay-mood-api 模块、ay-role-api 模块和接口模块创建完成之后，按照相同的步骤创建接口对应的服务模块：ay-user-service 模块、ay-mood-service 模块、ay-role-service 模块。接口模块和对应的服务模块创建完成之后，spring-boot-book-v2 的项目结构如图 16-9 所示。

为了方便，笔者把所有的模块都放在 spring-boot-book-v2 项目下，在真实的项目中并不是这样的。真实的项目中，会把接口和对应的服务单独建立一个项目，比如为 ay-user-api 和 ay-user-service 建立一个项目，ay-role-api 和 ay-role-service 建立一个项目，ay-mood-api 和 ay-mood-service 建立一个项目。这样不同的开发人员单独负责不同的项目，分工合作，提高开发效率。

所有的模块都建立好之后，可以把 spring-boot-book-v2 项目中的接口移动到对应的接口模块。比如把 spring-boot-book-v2 项目中的 AyUserService 接口移动到 ay-user-api，把 AyMoodService 接口移动到 ay-mood-api 等。同时把实现类 AyUserServiceImpl 移动到 ay-user-service，把实现类 AyMoodServiceImpl 移动到 ay-mood-service 等。这里以用户模块为例，讲解整个开发过程。

首先，ay-user-api 模块创建完成之后，该模块就是一个 spring-boot 微服务项目，享有 spring-boot 为我们默认生成的各种"福利"。我们在 ay-user-api 包下创建 api 包和 domain 包，分别用来存放接口类和实体类，在 api 包下存放所有 spring-boot-book-v2 项目移动过来的用户接口。ay-user-api 模块的目录如图 16-10 所示。

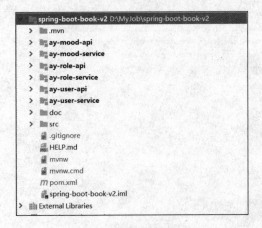

图16-9 spring-boot-book-v2 模块目录结构

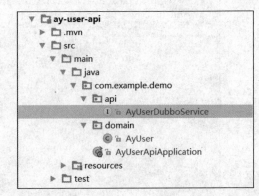

图16-10 ay-user-api 模块目录结构

在图 16-10 中，为了方便，在 api 包下创建 AyUserDubboService 接口，该接口用来提供与用户相关的服务，比如增删改查等功能。AyUserDubboService 类具体代码如下：

```
/**
 * 描述：用户接口
 * @author  Ay
 * @date    2017/12/16
 */
public interface AyUserDubboService {

    AyUser findByUserNameAndPassword(String name, String password);

}
```

在 AyUserDubboService 类中，只有一个通过用户名和密码查询用户的接口 findByUserNameAndPassword。domain 包下的 AyUser 类具体代码如下：

```
/**
 * 描述：用户实体类
 * @author  Ay
 * @date    2017/12/16
 */
public class AyUser {

    //主键
    private String id;
    //用户名
```

```
    private String name;
    //密码
    private String password;
    //邮箱
    private String mail;
    //省略 set、get 方法
}
```

这里还需要再重复强调一遍，api 包下的接口理论上是 spring-boot-book-v2 项目下的用户接口移动过来的，这里选择创建新的接口只是为了方便而已。

16.2.4 正式版发布

ay-user-api 模块接口开发完成之后，就可以将其发布成正式版 jar 包。jar 包有快照版 SNAPSHOT 和正式版。比如我们创建的 ay-user-api 模块默认就是快照版 0.0.1-SNAPSHOT。快照版的模块是可重复修改的，而正式版的模块不可修改。模块在开发过程中，基本都是 SNAPSHOT 版。当模块开发并测试完成会选择升级为正式版，而且在项目上线的时候，依赖的模块都必须是正式版，否则会出现意想不到的错误。我们可以在 pom 文件中修改模块的版本，比如把 ay-user-api 修改成 0.0.1 正式版，具体步骤如下：

步骤 01 在 ay-user-api 模块中，把 pom 文件 jar 包版本 version 升级为正式版 0.0.1，具体如图 16-11 所示。

步骤 02 版本修改之后，我们在 Intellij IDEA 开发工具右侧单击【clean】、【install】，将 ay-user-api 的 jar 包发布到本地 Maven 仓库，具体如图 16-12 所示。

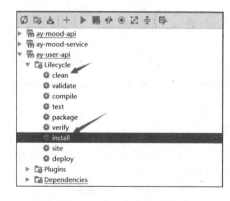

图 16-11 修改 pom 文件　　　　　　图 16-12 clean、install 命令窗口

步骤 03　ay-user-api 正式版的 jar 包发布成功之后，其他项目模块就可以使用它，同时我们可以到 Maven 仓库查看该 jar 包的具体信息，具体如图 16-13 所示。

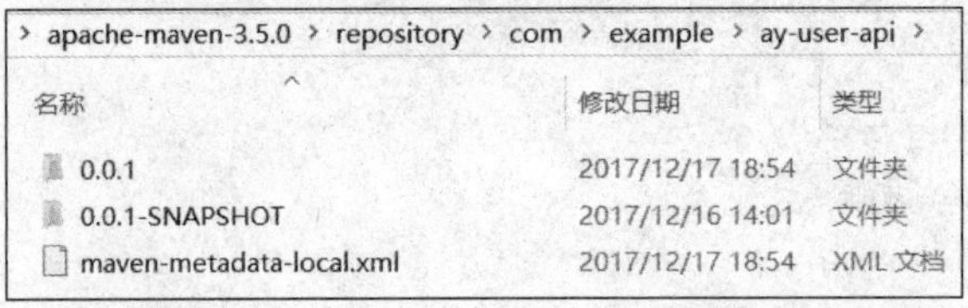

图 16-13　ay-user-api 的 maven 仓库信息

16.2.5　Service 服务端开发

ay-user-api 接口模块发布为正式版之后，我们继续开发 ay-user-service 模块。首先在 ay-user-service 模块中引入 ay-user-api 的依赖和 Dubbo 依赖。添加 ay-user-api 的依赖是因为 ay-user-service 模块是面向接口编程的，而添加 Dubbo 依赖是因为服务开发完成之后，要把该服务注册到 Zookeeper 中，具体代码如下：

```xml
<dependency>
    <groupId>com.example</groupId>
    <artifactId>ay-user-api</artifactId>
    <version>0.0.1</version>
</dependency>
<dependency>
    <groupId>io.dubbo.springboot</groupId>
    <artifactId>spring-boot-starter-dubbo</artifactId>
    <version>1.0.0</version>
</dependency>
```

在 ay-user-service 模块中添加完依赖之后，我们开发 AyUserDubboServiceImpl 服务来实现 ay-user-api 中的接口 AyUserDubboService，AyUserDubboServiceImpl 的具体代码如下：

```java
import com.alibaba.dubbo.config.annotation.Service;
import com.example.demo.api.AyUserDubboService;
import com.example.demo.domain.AyUser;
/**
 * 描述：对外提供用户服务类
 * Created by Ay on 2017/12/16
 */
```

```java
@Service(version = "1.0")
public class AyUserDubboServiceImpl implements AyUserDubboService {

    @Override
    public AyUser findByUserNameAndPassword(String name, String password) {
        //连接数据库，查询用户数据，此处省略
        AyUser ayUser = new AyUser();
        ayUser.setName("阿毅");
        ayUser.setPassword("123456");
        return ayUser;
    }
}
```

- @Service：这个注解不是Spring提供的，而是在com.alibaba.dubbo.config.annotation.Service包下面的，这一点要特别注意。在AyUserDubboServiceImpl类上添加@Service注解，就可以把AyUserDubboServiceImpl类注册到Zookeeper服务中心，对外提供服务了。version属性用于指定服务的版本，这里服务版本为1.0。当一个接口出现不兼容升级时，可以用版本号version过渡，版本号不同的服务相互间不引用。
- findByUserNameAndPassword方法：理论上应该在该方法中通过JPA Repository或者MyBatis查询用户数据，这里为了方便，简单创建AyUser对象返回。

16.2.6　Service 服务注册

ay-user-service 模块开发完成之后，由于 ay-user-service 本身是一个 Spring Boot 微服务项目，因此我们可以单独运行它。找到 ay-user-service 模块的入口类 AyUserServiceApplication（确保 Zookeeper 是启动状态），执行入口类的 main 方法便可启动 ay-user-service 服务。ay-user-service 服务启动完成之后，可以在 Intellij IDEA 的控制台中查看服务在 Zookeeper 的注册信息，具体信息如图 16-14 所示。

```
Session establishment complete on server 127.0.0.1/127.0.0.1:2181, sessionid = 0x16064c4eef00000, negotiated timeout = 
zookeeper state changed (SyncConnected)
 [DUBBO] Register: dubbo://192.168.141.1:20880/com.example.demo.api.AyUserDubboService?anyhost=true&application=provider
 [DUBBO] Subscribe: provider://192.168.141.1:20880/com.example.demo.api.AyUserDubboService?anyhost=true&application=pro
 [DUBBO] Notify urls for subscribe url provider://192.168.141.1:20880/com.example.demo.api.AyUserDubboService?anyhost=tr
Registering beans for JMX exposure on startup
Started AyUserServiceApplication in 3.673 seconds (JVM running for 5.405)
```

图 16-14　AyUserDubboService 注册信息

16.2.7　Client 客户端开发

ay-user-api 接口模块和 ay-user-service 服务模块开发完成之后，接下来开始开发客户端。所谓的客户端就是所有 ay-user-service 服务的对象。现在我们把 spring-boot-book-v2 作为客户端，spring-boot-book-v2 本身也是一个微服务。在 spring-boot-book-v2 项目中调用 ay-user-service 模块提供的服务，需要在 spring-boot-book-v2 项目的 pom 文件添加 ay-user-api 依赖和 Dubbo 依赖，具体代码如下：

```xml
<dependency>
    <groupId>com.example</groupId>
    <artifactId>ay-user-api</artifactId>
    <version>0.0.1</version>
</dependency>
<!-- dubbo start -->
<dependency>
    <groupId>io.dubbo.springboot</groupId>
    <artifactId>spring-boot-starter-dubbo</artifactId>
    <version>1.0.0</version>
</dependency>
```

ay-user-api 依赖和 Dubbo 依赖添加完成之后，就可以在代码中通过 @Reference 注解将 ay-user-service 模块提供的 AyUserDubboService 服务注入进来。具体代码如下：

```
@Reference(version = "1.0")
public AyUserDubboService ayUserDubboService;
```

@Reference 注解也是 Dubbo 框架提供的，在 com.alibaba.dubbo.config.annotation.Reference 包下。version 是 @Reference 注解的属性，version 的版本需要和 @Service 注解的 version 版本保持一致，否则服务将无法注入，这一点是需要特别注意的。

第 17 章

多环境配置与部署

本章主要介绍 Spring Boot 多环境配置及使用，Spring Boot 如何打成 war 包并部署到外部 Tomcat 服务器上等内容。

17.1 多环境配置概述

项目开发过程中，项目不同的角色会使用不同的环境。例如，开发人员会使用开发环境，测试人员会使用测试环境，性能测试会使用性能测试环境，项目开发完成之后会把项目部署到线上环境等。不同的环境往往会连接不同的 MySQL 数据库、Redis 缓存、MQ 消息中间件等，环境之间相互独立与隔离才不会相互影响，隔离的环境便于部署和提高工作效率，具体的环境隔离示意图如图 17-1 所示。

Spring Boot 中有两种配置文件，即 bootstrap（.yml 或者.properties）和 application（.yml 或者.properties）。为什么会有这两种配置文件呢？因为在 Spring Boot 中有两种上下文，一种是 bootstrap，另外一种是 application。boostrap 由父 ApplicationContext 加载，比 applicaton 优先加载。bootstrap 主要用于从额外的资源来加载配置信息，还可以在本地外部配置文件中解密属性。这两个上下文共用一个环境，它是任何 Spring 应用程序的外部属性的来源。bootstrap 里面的属性会优先加载，且不能被覆盖。

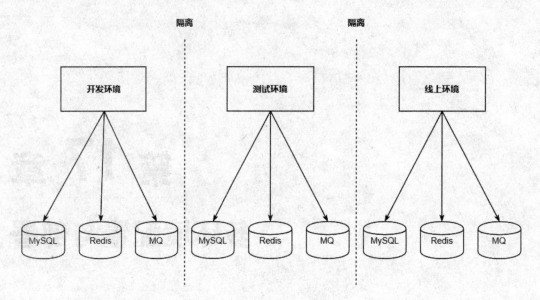

图 17-1 环境隔离

Spring Boot 支持通过外部配置覆盖默认配置项，具体优先级如下：

（1）Java 命令行参数。

（2）JNDI 属性。

（3）Java 系统属性（System.getProperties()）。

（4）操作系统环境变量。

（5）RandomValuePropertySource 配置的 random.*属性值。

（6）jar 包外部的 application-{profile}.properties 或 application.yml（带 spring.profile）配置文件。

（7）jar 包内部的 application-{profile}.properties 或 application.yml（带 spring.profile）配置文件。

（8）jar 包外部的 application.properties 或 application.yml（不带 spring.profile）配置文件。

（9）jar 包内部的 application.properties 或 application.yml（不带 spring.profile）配置文件。

（10）@Configuration 注解类上的@PropertySource。

（11）通过 SpringApplication.setDefaultProperties 指定的默认属性。

以 Java 命令行参数为例，运行 Spring Boot jar 包时，指定如下的参数：

```
### 参数用--xxx=xxx 的形式传递
java -jar app.jar --name=spring-boot --server.port=9090
```

应用启动的时候，就会覆盖默认的 Web Server 8080 端口，将其改为 9090。

17.2 多环境配置的使用

17.2.1 添加多个配置文件

假如项目 spring-boot-book-v2 需要 3 个环境：开发环境、测试环境、性能测试环境。我们复制 spring-boot-book-v2 项目配置文件 application.properties，分别取名为 application-dev.properties、application-test.properties、application-perform.properties，作为开发环境、测试环境、性能测试环境，具体如图 17-2 所示。

图 17-2　多环境配置文件

每个配置文件对应的 MySQL 数据库、Redis 缓存、ActiveMQ 消息队列等配置参数都不相同。

17.2.2 配置激活选项

多环境的配置文件开发完成之后，在 spring-boot-book-v2 的配置文件 application.properties 中添加配置激活选项，具体代码如下：

```
### 激活开发环境配置
spring.profiles.active=dev
```

如果想激活测试环境的配置，可修改为：

```
### 激活测试环境配置
spring.profiles.active=test
```

如果想激活性能测试环境的配置，可修改为：

```
### 激活性能测试环境配置
spring.profiles.active=perform
```

17.2.3 测试

多环境配置文件和配置激活选项开发完成之后，修改 application-dev.properties、application-test.properties、application-perform.properties 配置文件的数据库连接，具体代码如下所示。

开发环境配置文件 application-dev.properties，具体代码修改如下：

```
### 开发环境 MySQL 连接信息
spring.datasource.url=jdbc:mysql://127.0.0.1:3306/test?serverTimezone=UTC
```

测试环境配置文件 application-test.properties，具体代码修改如下：

```
### 测试环境 MySQL 连接信息
spring.datasource.url=jdbc:mysql://127.0.0.1:3306/test2?serverTimezone=UTC
```

性能测试环境配置文件 application-perform.properties，具体代码修改如下：

```
### 性能测试环境 MySQL 连接信息
spring.datasource.url=jdbc:mysql://127.0.0.1:3306/test3?serverTimezone=UTC
```

开发环境 MySQL 的 test 数据库已经存在，现在我们在 MySQL 数据库中创建 test2、test3 数据库作为测试环境和性能测试环境的数据库，并把 test 数据库中的数据导入到 test2、test3 数据库。我们可以利用第 2.3.3 节介绍的 Navicat for MySQL 客户端完成数据库数据的导入导出工作，具体步骤如下：

- 步骤 01　在 Navicat for MySQL 客户端中，找到 test 数据库，从鼠标右键菜单中选择【转储 SQL 文件】→【结构和数据】，将 test 数据库的数据存到指定的目录，如图 17-3 和图 17-4 所示。
- 步骤 02　test 数据库数据导出成功之后，在 MySQL 数据库中新建 test2 和 test3 数据库。
- 步骤 03　将步骤 1 中的数据导入到 test2 和 test3 数据库，从鼠标右键菜单中选择【运行 SQL 文件】，选择 test.sql 存放的目录，单击【开始】按钮进行数据导入，如图 17-5 和图 17-6 所示。

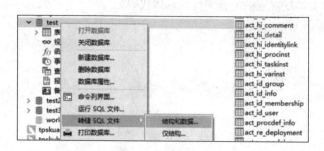

图 17-3　test 数据库导出操作　　　　图 17-4　数据库数据导出成功

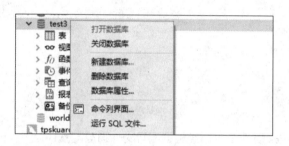

图 17-5　选择版本和组件　　　　　　图 17-6　选择版本和组件

测试环境数据库 test2 和性能测试环境 test3 数据库创建和数据导入成功之后，重新启动 spring-boot-book-v2 项目，项目成功启动之后，在浏览器输入访问地址：http://localhost:8080/ayUser/test，便可以成功访问到用户数据。如果想切换到测试环境进行项目开发，可以激活测试环境配置 spring.profiles.active=test，然后重新启动 spring-boot-book-v2t 项目即可。

17.3　自定义属性与加载

17.3.1　自定义属性

在使用 Spring Boot 的时候，通常需要自定义一些属性，可以按如下方式直接定义。在 src/main/resources/application.properties 配置文件中加入：

```
### 自定义属性
com.ay.book.name=spring boot 2
com.ay.book.author=ay
```

然后通过@Value("${属性名}")注解来加载对应的配置属性，具体代码如下：

```java
/**
 * 描述：自定义属性
 * @author Ay
 * @create 2019/08/31
 **/
@Component
public class BookProperties {

    @Value("${com.ay.book.name}")
    private String bookName;
    @Value("${com.ay.book.author}")
    private String author;

    //set、get 方法
}
```

最后，通过单元测试验证 BookProperties 属性是否已经根据配置文件加载配置：

```java
@RunWith(SpringRunner.class)
@SpringBootTest
public class DemoApplicationTests {

    @Resource
    private BookProperties bookProperties;

    @Test
    public void testProperties(){
        System.out.println(bookProperties.getBookName());
        System.out.println(bookProperties.getAuthor());
    }
}
```

不过我们并不推荐使用这种方式，下面给出更优雅的实现方式。
首先引入 Spring Boot 提供的配置依赖：

```xml
<!--spring boot 配置处理器 -->
<dependency>
```

```xml
    <groupId>org.springframework.boot</groupId>
    <artifactId>spring-boot-configuration-processor</artifactId>
    <optional>true</optional>
</dependency>
```

使用@ConfigurationProperties 注解进行编码，修改 BookProperties 为：

```java
/**
 * 描述：自定义配置（优雅实现）
 * @author Ay
 * @create 2019/08/31
 **/
@ConfigurationProperties(prefix="com.ay.book")
public class BookProperties {

    private String name;

    private String author;

    //省略 set、get 方法
}
```

- @ConfigurationProperties(prefix="com.ay.book")：在application.properties配置的属性前缀。在类中的属性就不用使用@value进行注入了。

最后，在启动类中添加@EnableConfigurationProperties({BookProperties.class})。

17.3.2　参数间的引用

在 application.properties 中的各个参数之间也可以直接引用来使用，例如下面的设置：

```
### 自定义属性
com.ay.book.name=spring boot 2
com.ay.book.author=ay
### 引用 com.ay.book.name 和 com.ay.book.author 配置
com.ay.book.desc=${com.ay.book.name}: ${com.ay.book.author}
```

com.ay.book.desc 参数引用上下文定义的 name 和 author 属性，最后修改属性的值为：

```
spring boot 2: ay
```

17.3.3 使用随机数

Spring Boot 的属性配置文件中可以通过${random}来产生 int 值、long 值、string 字符串或者 UUID，来支持属性的随机值。按以下方法从配置文件中获取符合规则的随机数：

```
### 随机字符串：32位MD5字符串
com.ay.book.randomValue=${random.value}
### 随机生成int类型的数
com.ay.book.randomInt=${random.int}
### 随机生成long类型的数
com.ay.book.randomLong=${random.long}
### 随机生成uuid
com.ay.book.randomUuid=${random.uuid}
### 10以内的随机数
com.ay.book.randomLen=${random.int(10)}
### 随机生成[1024,65536]范围内的数
com.ay.book.randomRange=${random.int[1024,65536]}
```

将 BookProperties.java 类的代码修改如下：

```
/**
 * 描述：自定义配置（优雅实现）
 * @author Ay
 * @create 2019/08/31
 **/
@ConfigurationProperties(prefix="com.ay.book")
public class BookProperties {

    private String name;

    private String author;

    private String desc;

    private String randomValue;

    private Integer randomInt;

    private Long randomLong;
```

```java
    private String randomUuid;

    private Integer randomLen;

    private Integer randomRange;

    //省略 set、get 方法
}
```

在测试类中测试,具体代码如下:

```java
@RunWith(SpringRunner.class)
@SpringBootTest
public class DemoApplicationTests {

    @Resource
    private BookProperties2 bookProperties2;

    @Test
    public void contextLoads() {
        System.out.println("value:" + bookProperties2.getRandomValue());
        System.out.println("int:" + bookProperties2.getRandomInt());
        System.out.println("long:" + bookProperties2.getRandomLong());
        System.out.println("uuid:" + bookProperties2.getRandomUuid());
        System.out.println("len" + bookProperties2.getRandomLen());
        System.out.println("range:" + bookProperties2.getRandomRange());
    }

}
```

输出结果如下:

```
value:3a4f8ede4cc2a283fe39ae46ae9ceaac
int:-713659987
long:3818318578637578592
uuid:92eba27f-34a1-4c7c-86f9-3128276ce498
len6
range:24879
```

17.4 部署

17.4.1 Spring Boot 内置 Tomcat

Tomcat 是一个免费的开放源代码的 Web 应用服务器，属于轻量级应用服务器。在中小型系统和并发访问用户不是很多的场合下被普遍使用。Spring Boot 默认使用 Tomcat 作为内嵌 Servlet 容器，查看 spring-boot-starter-web 依赖可知，如图 17-7 所示。本书使用 Tomcat 8.0 进行讲解，可到官方网站 https://tomcat.apache.org/download-80.cgi 进行下载，下载完成之后解压到 D 盘，并将解压后的文件夹命名为 tomcat8，如图 17-8 所示。

图 17-7 Tomcat 解压目录

图 17-8 文件夹命名为 tomcat8

如果想用使用其他 servlet 容器，比如 Jetty 作为 Spring Boot 默认内置容器，只需要修改 spring-boot-starter-web 依赖即可。使用 Jetty 容器作为 Spring Boot 默认内置容器，具体修改代码如下：

```xml
<dependency>
    <groupId>org.springframework.boot</groupId>
    <artifactId>spring-boot-starter-web</artifactId>
    <exclusions>
        <exclusion>
            <groupId>org.springframework.boot</groupId>
            <artifactId>spring-boot-starter-tomcat</artifactId>
        </exclusion>
    </exclusions>
</dependency>
<dependency>
    <groupId>org.springframework.boot</groupId>
    <artifactId>spring-boot-starter-jetty</artifactId>
</dependency>
```

17.4.2　Intellij IDEA 配置 Tomcat

在 Intellij IDEA 配置 Tomcat，具体步骤如下：

步骤 01　在 IDEA 开发菜单栏中，选择【run】→【Edit Configurations】，在弹出的窗口中选择【Defaults】→【Tomcat Server】→【Local】，在【Application server】中选择 Tomcat 的安装路径，在【JRE】中选择 JDK 的安装路径，具体如图 17-9 所示。

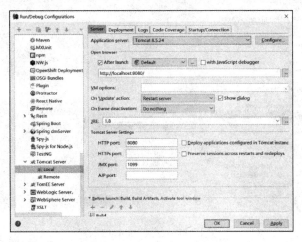

图 17-9　Tomcat 配置

步骤 02　在【Deployment】选项中，选择【Artifact】→【spring-boot-book-v2:war exploded】，如图 17-10 所示。

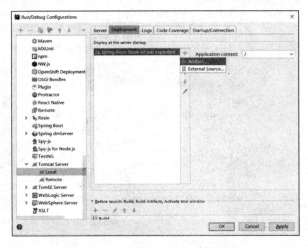

图 17-10　配置 war 包

步骤03 步骤1和步骤2只是配置一个Defaults默认Tomcat模板,现在我们单击【+】加号按钮→【Tomcat Server】→【Local】,在弹出的界面中输入Name为tomcat8,其他信息会从默认模板中获取到,具体如图17-11和图17-12所示。

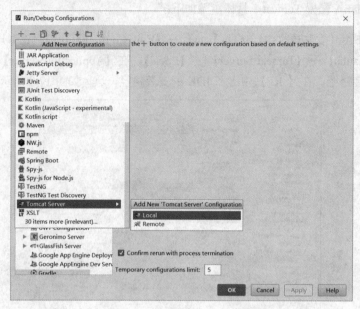

图17-11 创建tomcat配置

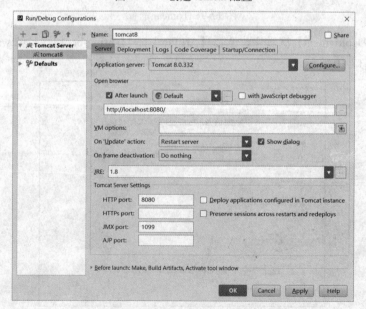

图17-12 修改tomcat名称

17.4.3　war 包部署

在 IDEA 开发工具中配置完 Tomcat 之后，修改 spring-boot-book-v2 项目的 pom.xml 文件，将配置：

```
<packaging>jar</packaging>
```

修改为

```
<packaging>war</packaging>
```

配置修改完成之后，使用 maven clean、maven package 和 maven install 命令，具体如图 17-13 所示。将项目重新打包。此时就可以在 spring-boot-book-v2 项目下看到 war 包，具体如图 17-14 所示。

图 17-13　Maven 命令

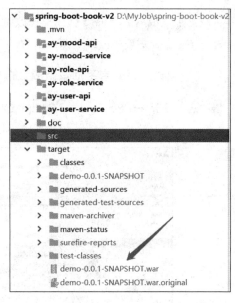

图 17-14　target 下 war 包

17.4.4　测试

demo-0.0.1-SNAPSHOT.war 包生成之后，就可以启动 spring-boot-book-v2 项目。现在不是使用项目入口类 DemoApplication 启动项目，而是将 spring-boot-book-v2 以 war 包的方式部署到外置的 tomcat 服务器上。tomcat 启动成功之后，就可以在浏览器上访问 spring-boot-book-v2 项目。

17.5 热部署

在 Spring Boot 中可以很方便地实现代码热部署，代码的修改可以自动部署并重新热启动项目。具体的方法是，在 pom.xml 配置文件中添加 devtools 依赖。

```
<!-- 热部署 -->
<dependency>
        <groupId>org.springframework.boot</groupId>
        <artifactId>spring-boot-devtools</artifactId>
        <optional>true</optional>
</dependency>
```

依赖添加完成后，当 Java 代码类修改时就会热更新。

在 IDEA 中，需要修改以下两个地方。

1. 勾选自动编译或者手动重新编译选项

依次选择 File → Settings → Compiler，并勾选 Build project automatically 选项，如图 17-15 所示。

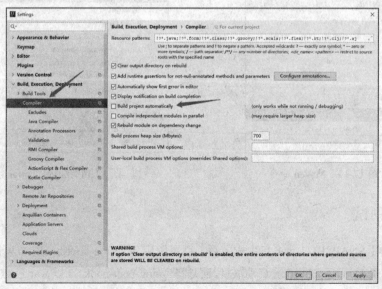

图 17-15　勾选 Build project automatically 选项

2. 组合键【Shift+Ctrl+Alt+/】选择【Registry】，勾选【compiler.automake.allow.when.
app.running】

需要注意的是，生产环境 devtools 将被禁用，如 java -jar 方式或者自定义的类加载器等都会识别为生产环境。应用打包默认也不会包含 devtools，除非你禁用 Spring Boot Maven 插件的 excludeDevtools 属性。

17.6 思考题

1. 什么是嵌入式服务器？为什么要使用嵌入式服务器？

答：思考一下我们在虚拟机上部署应用程序时，需要哪些步骤：

（1）安装 Java。
（2）安装 Web 或者应用程序的服务器（Tomat/Jetty 等）。
（3）部署应用程序 war 包。

如果想简化这些步骤 —— 这个想法是嵌入式服务器的起源。当创建一个可以部署的应用程序时，将会把服务器（例如 tomcat）嵌入到可部署的服务器中。例如，对于一个 Spring Boot 应用程序来说，可以生成一个包含 Embedded Tomcat 的应用程序 jar。此时，就可以像运行正常 Java 应用程序一样来运行 Web 应用程序了。嵌入式服务器就是可执行单元包含服务器的二进制文件（例如 tomcat.jar）。

2. Spring Boot 的核心注解是哪个？它主要由哪几个注解组成？

答：启动类上面的注解是@SpringBootApplication，它也是 Spring Boot 的核心注解，主要包含了以下 3 个注解。

- @SpringBootConfiguration：组合了@Configuration 注解，用于实现配置文件的功能。
- @EnableAutoConfiguration：打开自动配置的功能，也可以关闭某个自动配置的选项，如关闭数据源自动配置功能：@SpringBootApplication(exclude = { DataSourceAutoConfiguration.class })。
- @ComponentScan：Spring组件扫描。

3. Spring Boot 需要独立的容器运行吗？

答：可以不需要，因为已内置了 Tomcat/ Jetty 等容器。

4. Spring Boot 有哪几种读取配置的方式？

答：Spring Boot 可以通过 @PropertySource、@Value 和 @Environment、@ConfigurationProperties 来绑定变量。

5. Spring Boot 实现热部署有哪几种方式？

答：主要有两种方式，即 Spring Loaded 和 Spring-boot-devtools 方式。

6. Spring Boot 如何定义多套不同的环境配置？

答：提供多套配置文件，如 applcation.properties、application-dev.properties、application-test.properties 等。

第 18 章

微服务容器化

本章主要介绍 Docker 技术以及如何将 Spring Boot 项目容器化等内容。

18.1 Docker 概述

18.1.1 Docker 的优势

Docker 是一个开源的应用容器引擎,基于 Go 语言并遵从 Apache 2.0 协议开源。Docker 可以让开发者打包他们的应用以及依赖包到一个轻量级、可移植的容器中,然后发布到任何流行的 Linux 机器上,也可以实现虚拟化。容器完全使用沙箱机制,相互之间不会有任何接口,更重要的是容器性能开销极低。

作为一种新兴的虚拟化方式,Docker 与传统的虚拟化方式相比具有众多的优势。

(1) 高效的利用系统资源

由于容器不需要进行硬件虚拟以及运行完整操作系统等额外开销,Docker 对系统资源的利用率更高。无论是应用的执行速度、内存损耗或者文件存储速度,都要比传统虚拟机技术更高效。因此,相比虚拟机技术,一个相同配置的主机往往可以运行更多数量的应用。

(2) 快速的启动时间

传统的虚拟机技术启动应用服务往往需要数分钟,而 Docker 容器应用由于直接运行于宿主内核，无须启动完整的操作系统，因此可以做到秒级甚至毫秒级的启动时间，大大地节约了开发、测试、部署的时间。

（3）一致的运行环境

开发过程中一个常见的问题是环境一致性问题。由于开发环境、测试环境、生产环境不一致，导致有些bug并未在开发过程中被发现。而Docker的镜像提供了除内核外完整的运行时环境，确保了应用运行环境一致性，从而不会再出现"这段代码在我机器上没问题啊"这类问题。

（4）持续交付和部署

对开发和运维人员来说，最希望的就是一次创建或配置，可以在任意地方正常运行。使用Docker 可以通过定制应用镜像来实现持续集成（Continuous Integration）、持续交付和部署（Continuous Delivery/Deployment）。开发人员可以通过 Dockerfile 来进行镜像构建，并结合持续集成系统进行集成测试，而运维人员则可以直接在生产环境中快速部署该镜像，甚至结合持续部署系统进行自动部署。而且使用 Dockerfile 使镜像构建透明化，不仅仅开发团队可以理解应用运行环境，也方便运维团队理解应用运行所需的条件，帮助更好地在生产环境中部署该镜像。

（5）迁移简单

由于 Docker 确保了执行环境的一致性，使得应用的迁移更加容易。Docker 可以在很多平台上运行，无论是物理机、虚拟机、公有云、私有云，甚至是笔记本上，其运行结果是一致的。因此用户可以很轻易地将在一个平台上运行的应用迁移到另一个平台上，而不用担心运行环境的变化导致应用无法正常运行的情况。

（6）容易维护和扩展

当我们需要在宿主机器上运行一个虚拟操作系统时，往往需要安装虚拟软件，如 Oracle VirtualBox 或者 VMware。当我们在虚拟软件上安装操作系统的时候，虚拟机软件需要模拟 CPU、内存、I/O 设备和网络资源等，为了能运行应用程序，除了需要部署应用程序本身及其依赖，还需要安装整个操作系统和驱动，会占用大量的系统开销。

Docker 使用的分层存储及镜像技术，使得应用重复部分的复用更为容易，也使得应用的维护更新更加简单，基于基础镜像进一步扩展镜像也变得非常简单。此外，Docker 团队同各个开源项目团队一起维护了一大批高质量的官方镜像，既可以直接在生产环境使用，又可以作为基础进一步定制，大大降低了应用服务镜像的制作成本。

下面我们来简单对比一下传统虚拟机和 Docker。具体见表18-1。

表 18-1 Docker 容器与虚拟机对比

特　　性	Docker 容器	虚　拟　机
性能	接近原生	弱于原生
启动	秒级	分钟级
硬盘使用	一般为 MB	一般为 GB
系统支持量	单机支持上千个容器	一般几十个

从表中的数据来看，虚拟机和 Docker 容器虽然可以提供相同的功能，但是优缺点是显而易见的。

18.1.2 Docker 的基本概念

下面我们来理解一下 Docker 的几个基本概念：镜像（Image）、容器（Container）、仓库（Respository）与镜像注册中心（Docker Registry）。

1. 镜像（Image）

Docker 镜像可以理解为一个 Linux 文件系统，Docker 镜像是一个特殊的文件系统，除了提供容器运行时所需的程序、库、资源、配置等文件外，还包含了一些为运行时准备的一些配置参数（如匿名卷、环境变量、用户等）。

2. 容器（Container）

镜像（Image）和容器（Container）的关系，就像是面向对象程序设计中的类和实例一样，镜像是静态的定义，容器是镜像运行时的实体。容器可以被创建、启动、停止、删除、暂停等。

容器可以拥有自己的 root 文件系统、自己的网络配置、自己的进程空间，甚至自己的用户 ID 空间。容器内的进程运行在一个隔离的环境里，使用起来就好像是在一个独立于宿主的系统下操作一样。这种特性使得容器封装的应用比直接在宿主机上运行更加安全。

3. 仓库（Respository）与镜像注册中心（Docker Registry）

镜像构建完成后，如果需要在其他服务器上使用这个镜像，就需要一个集中的存储、分发镜像的服务，Docker Registry 镜像注册中心就是这样的服务。

一个 Docker Registry 镜像注册中心可以包含多个仓库（Repository），每个仓库可以包含多个标签（Tag），每个标签对应一个镜像。通常，一个仓库会包含同一个软件不同版本的镜像，而标签就常用于对应该软件的各个版本。我们可以通过<仓库名>:<标签>的格式来指定具体是这个软件哪个版本的镜像。如果不给出标签，将以 latest 作为默认标签。

Docker 官方提供了一个叫作 Docker Hub 的镜像注册中心，用于存放公有和私有的 Docker 镜像仓库（Repository）。可以通过 Docker Hub 下载 Docker 镜像，也可以将自己创建的 Docker 镜像上传到 Docker Hub 上。Docker Hub 的网址为：https://hub.docker.com/。

18.1.3　Docker 架构

我们先来了解一下 Docker 引擎。Docker 引擎可以理解为一个运行在服务器上的后台进程，其主要包括 3 大组件。

- Docker 后台服务（Docker Daemon）：长时间运行在后台的守护进程，是 Docker 的核心服务，可以通过 dockerd 命令与它交互通信。
- REST 接口（REST API）：程序可以通 REST 的接口来访问后台服务，或向它发送操作指令。
- 交互式命令行界面（Docker CLI）：可以使用命令行界面与 Docker 进行交互，例如以 docker 为开头的所有命令的操作。而命令行界面又是通过调用 REST 的接口来控制和操作 Docker 后台服务的，如图 18-1 所示。

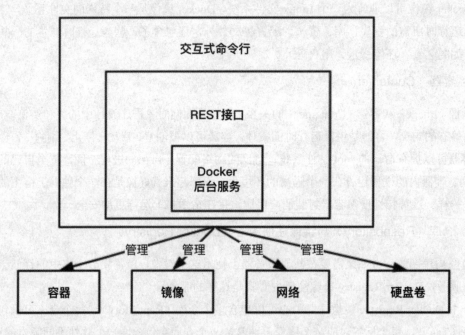

图 18-1　Docker 引擎 3 大组件架构图

我们再来看看 Docker 官方文档上的一张架构图，具体如图 18-2 所示。

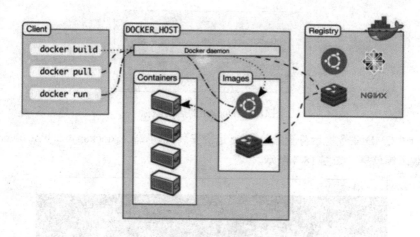

图 18-2 Docker 系统架构

- Docker客户端（Docker Client）：与Docker后台服务交互的主要工具，在使用docker run命令时，客户端把命令发送到Docker后台服务，再由后台服务执行该命令。可以使用docker build命令创建Docker镜像，使用docker pull命令拉起Docker镜像，使用docker run命令运行Docker镜像，从而启动Docker容器。
- Docker Host：表示运行Docker引擎的宿主机，包括Docker daemon后台进程，可通过该进程创建Docker镜像，并在Docker镜像上启动Docker容器。
- Registry：表示Docker官方镜像注册中心，包含大量的Docker镜像仓库，可以通过Docker引擎拉取所需的Docker镜像到宿主机上。

18.1.4　Docker 的安装

Docker 分为两个版本：社区版（Community Edition，CE）和企业版（Enterprise Edition，EE）。Docker 社区版主要提供给开发者学习和练习，而企业版主要提供给企业级开发和运维团队用于对线上产品的编译、打包和运行，有很高的安全性和扩展性。Docker 的社区版和企业版都支持 Linux、Cloud、Windows 和 Mac OS 平台等，这里我们以 Windows 操作系统为例演示 Docker 的安装，具体步骤如下：

步骤01 Windows 操作系统可以利用 docker toolbox 来安装，国内可以使用阿里云的镜像下载，下载地址：http://mirrors.aliyun.com/docker-toolbox/windows/docker-toolbox/，本书选择DockerToolbox-18.03.0-ce.exe 版本。

步骤02 用鼠标双击 DockerToolbox-18.03.0-ce.exe，按照提示一步一步安装即可，安装完成后在桌面上会出现两个图标，一个是命令行形式的 Docker 终端，另一个是图形界面的 Docker 操作工具，如图 18-3 所示。

图 18-3 Docker 的启动图标

步骤 03 通过 CMD 命令提示符查看 Docker 是否安装成功,输入 docker -machine 会出现 Docker 版本等信息,如图 18-4 所示。

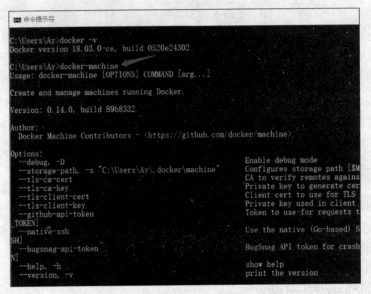

图 18-4 查看 Docker 是否安装成功

步骤 04 将 C:\Program Files\Docker Toolbox 文件夹下的 boot2docker.iso 复制到 C:\Users\ay\.docker\machine\cache 下,然后断开网络。

步骤 05 双击桌面的 Docker Quickstart Terminal 终端,启动命令行终端,该过程比较缓慢,需要耐心等待。当出现如图 18-5 所示的界面,代表命令行终端启动成功。

在命令行终端输入命令"docker run hello-world",报请求超时异常。

```
$ docker run hello-world
Unable to find image 'hello-world:latest' locally
C:\Program Files\Docker Toolbox\docker.exe: Error response from daemon: Get
https://registry-1.docker.io/v2/: net/http: request canceled (Client.Timeout
exceeded while awaiting headers).
See 'C:\Program Files\Docker Toolbox\docker.exe run --help'.
```

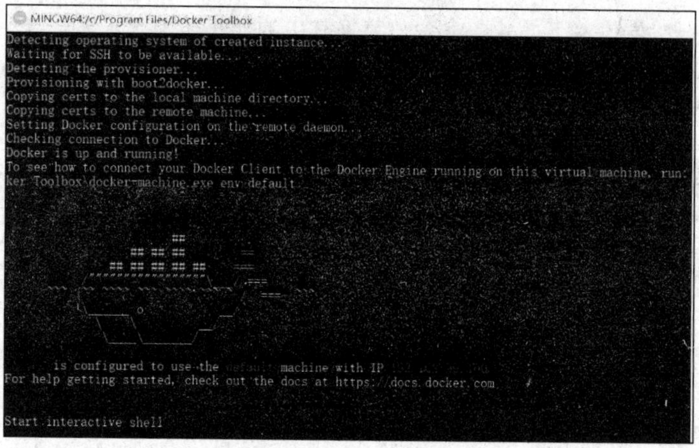

图 18-5 启动 Docker 命令行终端

原因是镜像都在国外，访问速度慢，所以我们需要换成国内的下载源，具体更改步骤如下：

步骤 01 访问 DaoCloud 网站：https://www.daocloud.io/mirror#accelerator-doc 并注册账号，可以使用 Git hub 账号登录。注册并登录成功后，单击加速按钮，在页面的最后面找到加速地址，具体如图 18-6 和图 18-7 所示。

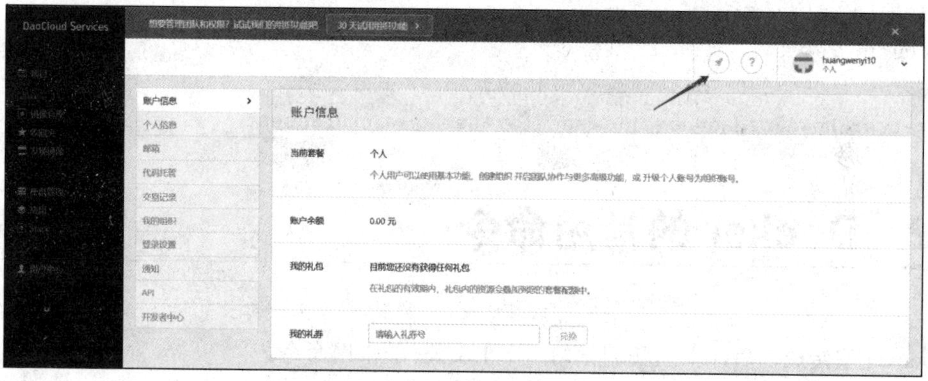

图 18-6 DaoCloud 首页

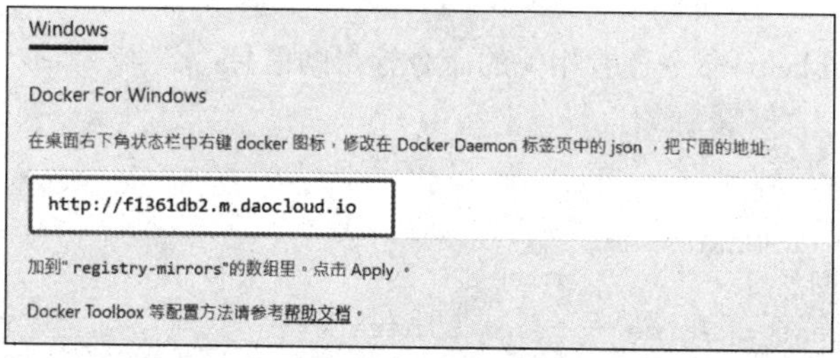

图 18-7 加速地址

步骤 02 在命令行终端输入下面 4 条命令（替换其中的加速地址）。

```
### ssh 到 machine
docker-machine ssh default
### 替换加速地址，读者可自学 sed 命令，sed 命令是工作中非常常用的命令，
### 例如：sed -i 's/原字符串/新字符串/g' ab.txt 表示对全局匹配上的所有字符串进行替换
sudo sed -i "s|EXTRA_ARGS='|EXTRA_ARGS='--registry-mirror=加速地址 |g" /var/lib/boot2docker/profile
### 退出
exit
### 重启
docker-machine restart default
```

步骤 03 在命令行终端输入命令"docker run hello-world"，便可以下载 hello-world 镜像，并运行它。

```
$ docker run hello-world
Unable to find image 'hello-world:latest' locally
latest: Pulling from library/hello-world
1b930d010525: Pull complete
Digest: sha256:6540fc08ee6e6b7b63468dc3317e3303aae178cb8a45ed3123180328bcc1d20f
Status: Downloaded newer image for hello-world:latest
```

18.2 Docker 的常用命令

Docker 安装成功后，下面我们学习一下 Docker 的常用命令。

1. 查看版本信息

（1）使用 docker help 命令查看相关的命令行帮助信息：

```
$ docker help

Usage:  docker COMMAND

A self-sufficient runtime for containers

Options:
```

```
      --config string      Location of client config files (default
                           "C:\\Users\\Ay\\.docker")
  -D, --debug              Enable debug mode
  -H, --host list          Daemon socket(s) to connect to
  -l, --log-level string   Set the logging level
                           ("debug"|"info"|"warn"|"error"|"fatal")
                           (default "info")
      --tls                Use TLS; implied by --tlsverify
      --tlscacert string   Trust certs signed only by this CA (default
                           "C:\\Users\\Ay\\.docker\\machine\\machines\\default\\ca.pem")
      --tlscert string     Path to TLS certificate file (default
                           "C:\\Users\\Ay\\.docker\\machine\\machines\\default\\cert.pem")
      --tlskey string      Path to TLS key file (default
                           "C:\\Users\\Ay\\.docker\\machine\\machines\\default\\key.pem")
      --tlsverify          Use TLS and verify the remote (default true)
  -v, --version            Print version information and quit
```

（2）使用 docker -v 查看 Docker 的版本信息：

```
$ docker -v
Docker version 18.03.0-ce, build 0520e24302
```

2. 镜像相关命令

（1）使用 docker images 查看本地镜像：

```
$ docker images
REPOSITORY        TAG         IMAGE ID          CREATED           SIZE
hello-world       latest      fce289e99eb9      6 months ago      1.84kB
```

- REPOSITORY：表示镜像仓库名称。
- TAG：表示标签名称，latest 表示最新版本。
- IMAGE_ID：镜像 ID，唯一的，这里我们可以看到 12 位字符串，实际上它是 64 位完整镜像 ID 的缩略表达式。
- CREATE：表示镜像的创建时间。
- SIZE：表示镜像的字节大小。

（2）如果想拉取 Docker Hub 中的镜像，可以使用 docker pull 命令：

```
### 从 Docker Hub 拉取 tomcat 镜像
$ docker pull tomcat
Using default tag: latest
latest: Pulling from library/tomcat
a4d8138d0f6b: Pull complete
dbdc36973392: Pull complete
f59d6d019dd5: Pull complete
aaef3e026258: Pull complete
5e86b04a4500: Pull complete
1a6643a2873a: Pull complete
2ad1e30fc17c: Pull complete
16f4e6ee0ca6: Pull complete
928f4d662d23: Pull complete
b8d24294d525: Pull complete
Digest: sha256:bace3e146f40bdc1ab0d781cefe51271fd61e532550b39e86d67ec710eafa45e
Status: Downloaded newer image for tomcat:latest
```

（3）如果想在 Docker Hub 中搜索镜像，可以使用 docker search 命令：

```
### 搜索 tomcat 镜像
$ docker search tomcat
NAME             DESCRIPTION                                    STARS    OFFICIAL   AUTOMATED
Tomcat           Apache Tomcat is an open source implementati…   2464               [OK]
Tome             Apache TomEE is an all-Apache Java EE certif…   66                 [OK]
dordoka/tomcat   Ubuntu 14.04, Oracle JDK 8 and Tomcat 8 base…   53                 [OK]
bitnami/tomcat   Bitnami Tomcat Docker Image                     29
```

- **NAME**：表示镜像名称。
- **DESCRIPTION**：表示镜像仓库的描述。
- **STARS**：表示镜像仓库的收藏数，用户可以在 Docker Hub 上对镜像仓库进行收藏，一般可以通过收藏数判断该镜像的受欢迎程度。
- **OFFICIAL**：表示是否为官方仓库，官方仓库具有更高的安全性。
- **AUTOMATED**：表示是否自动构建镜像仓库，用户可以将自己的 Docker Hub 绑定到 Github 账号上，当代码提交后，可自动构建镜像仓库。

（4）如果想导入/导出镜像，可以使用 docker save 或者 docker load 命令

```
### 导出 centos 镜像为一个 tar 文件，若不指定导出的 tar 路径，默认为当前目录
$ docker save centos > centos.jar
```

导出的 centos 镜像包文件可随时在另一台 Docker 机器导入，命令如下：

```
$ docker load < centos.jar
```

（5）如果想删除镜像，可使用 docker rmi 命令

```
### 查询 docker 镜像
$ docker images
REPOSITORY          TAG         IMAGE ID        CREATED         SIZE
tomcat              latest      238e6d7313e3    2 days ago      506MB
hello-world         latest      fce289e99eb9    6 months ago    1.84kB
### 删除镜像，fce289e99eb9 为镜像 ID
$ docker rmi fce289e99eb9
Error response from daemon: conflict: unable to delete fce289e99eb9 (must be forced) - image is being used by stopped container fc3477d35bba
```

3. 容器相关命令

（1）如果想查看运行的容器，可使用如下命令：

```
### 查看运行的容器，由于没启动任何容器，故没有相关的记录
$ docker ps
CONTAINER ID    IMAGE       COMMAND         CREATED         STATUS      PORTS       NAMES
### 启动 tomcat 容器
$ docker run tomcat
20-Jul-2019 15:27:05.995 INFO [main] org.apache.catalina.startup.VersionLoggerListener.log Server version:        Apache Tomcat/8.5.43
20-Jul-2019 15:27:06.022 INFO [main] org.apache.catalina.startup.VersionLoggerListener.log Server built:          Jul 4 2019 20:53:15 UTC
//省略代码
### 再次运行 docker ps 命令，便可以看出目前运行的容器：6a25e44d102d
$ docker ps
CONTAINER ID    IMAGE       COMMAND             CREATED         STATUS          PORTS       NAMES
6a25e44d102d    tomcat      "catalina.sh run"   36 seconds ago  Up 34 seconds   8080/tcp    awesome_edison
```

列出最近创建的容器，包括所有的状态
```
$ docker ps -l
CONTAINER ID        IMAGE              COMMAND             CREATED           STATUS           PORTS        NAMES
6a25e44d102d        tomcat             "catalina.sh run"   9 minutes ago     Up 9 minutes     8080/tcp     awesome_edison
```
列出所有的容器，包括所有的状态
```
$ docker ps -a
CONTAINER ID        IMAGE              COMMAND             CREATED           STATUS              PORTS        NAMES
6a25e44d102d        tomcat             "catalina.sh run"   11 minutes ago    Up 11 minutes       8080/tcp     awesome_edison
5271c2510128        hello-world        "/hello"            11 minutes ago    Exited (0) 11 minutes ago        vigilant_bohr
fc3477d35bba        hello-world        "/hello"            2 hours ago       Exited (0) 2 hours ago           reverent_wright
```
-q 表示仅列出 CONTAINER ID
```
$ docker ps -a -q
6a25e44d102d
5271c2510128
fc3477d35bba
```
可以使用 help 命令查看更多关于 docker ps 用法，其他命令类似
```
$ docker ps --help

Usage: docker ps [OPTIONS]

List containers

Options:
  -a, --all             Show all containers (default shows just running)
  -f, --filter filter   Filter output based on conditions provided
      --format string   Pretty-print containers using a Go template
  -n, --last int        Show n last created containers (includes all
                        states) (default -1)
  -l, --latest          Show the latest created container (includes all
                        states)
      --no-trunc        Don't truncate output
  -q, --quiet           Only display numeric IDs
  -s, --size            Display total file sizes
```

- CONTAINER ID：表示容器ID。
- IMAGE：表示镜像名称。
- COMMAND：表示启动容器运行的命令，Docker容器要求我们在启动容器时需要运行一个命令。
- CREATE：表示容器创建的时间。
- STATUS：表示容器运行的状态，例如UP表示运行中，Exited表示已退出。
- PORTS：表示容器需要对外暴露的端口。
- NAMES：表示容器的名称，由Docker引擎自动生成，也可以在docker run命令中通过--name选项来指定。

（2）如果想创建并启动容器，可以使用如下命令：

```
### 启动centos容器，并进入到容器里
$ docker run -i -t centos /bin/bash
Unable to find image 'centos:latest' locally
latest: Pulling from library/centos
8ba884070f61: Pull complete
Digest: sha256:b40cee82d6f98a785b6ae35748c958804621dc0f2194759a2b8911744457337d
Status: Downloaded newer image for centos:latest
[root@9114b798696b /]#
```

- -i选项：表示启动容器后，打开标准输入设备（STDIN），可以使用键盘进行输入。
- -t选项：表示启动容器后，分配一个伪终端，将与容器建立会话。
- centos参数：表示要运行的镜像名称，标准格式为：centos:latest，若为latest版本，可省略latest。
- /bin/bash参数：表示运行容器中的bash应用程序。

 上述命令首先从本地获得CentOs镜像，若本地没有此镜像，则从Docker Hub上拉取CentOs镜像并放入本地，随后根据CentOs镜像创建并启动CentOs容器。

除了使用该命令可以创建和进入容器外，还可以使用如下命令进入运行中的容器：

```
### 进入启动中的容器，但是不能进入已停止的容器
$ docker attach 9114b798696b
### root表示以超级管理员身份进入容器，9114b798696b表示容器的ID，/表示当前路径
[root@9114b798696b /]#
```

还可以使用以下命令向运行中的容器执行具体命令：

```
### 9114b798696b 容器 ID，ls -l 表示列出容器中当前的目录结构
$ docker exec -i -t centos ls -l
total 56
-rw-r--r--   1 root root 12082 Mar  5 17:36 anaconda-post.log
lrwxrwxrwx   1 root root     7 Mar  5 17:34 bin -> usr/bin
drwxr-xr-x   5 root root   360 Jul 21 06:00 dev
drwxr-xr-x  47 root root  4096 Jul 21 06:00 etc
drwxr-xr-x   2 root root  4096 Apr 11  2018 home
lrwxrwxrwx   1 root root     7 Mar  5 17:34 lib -> usr/lib
lrwxrwxrwx   1 root root     9 Mar  5 17:34 lib64 -> usr/lib64
drwxr-xr-x   2 root root  4096 Apr 11  2018 media
drwxr-xr-x   2 root root  4096 Apr 11  2018 mnt
drwxr-xr-x   2 root root  4096 Apr 11  2018 opt
dr-xr-xr-x 156 root root     0 Jul 21 06:00 proc
dr-xr-x---   2 root root  4096 Mar  5 17:36 root
drwxr-xr-x  11 root root  4096 Mar  5 17:36 run
lrwxrwxrwx   1 root root     8 Mar  5 17:34 sbin -> usr/sbin
drwxr-xr-x   2 root root  4096 Apr 11  2018 srv
dr-xr-xr-x  13 root root     0 Jul 21 05:54 sys
drwxrwxrwt   7 root root  4096 Mar  5 17:36 tmp
drwxr-xr-x  13 root root  4096 Mar  5 17:34 usr
drwxr-xr-x  18 root root  4096 Mar  5 17:34 var
```

可以使用 docker stop 和 docker kill 命令停止或者终止容器：

```
### 停止运行中的容器，9114b798696b 为容器 ID
$ docker stop 9114b798696b
9114b798696b
### 再次创建和启动容器
$ docker run -i -t centos /bin/bash
[root@1dc3f698743e /]#
### 终止运行中的容器，1dc3f698743e 为容器 ID
➜    docker kill 1dc3f698743e
1dc3f698743e
```

可以使用 docker start 和 docker restart 命令启动或者重启容器：

```
### 启动已停止的容器，1dc3f698743e 为容器 ID
$ docker start 1dc3f698743e
1dc3f698743e
### 重启运行中的容器，1dc3f698743e 为容器 ID
$ docker restart 1dc3f698743e
1dc3f698743e
```

可以使用 docker rm 命令来删除已经停止的容器：

```
### 停止容器 ID 为 1dc3f698743e 的容器
$ docker stop 1dc3f698743e
1dc3f698743e
### 删除已停止的容器，1dc3f698743e 为容器 ID
$ docker rm 1dc3f698743e
1dc3f698743e
###强制删除所有运行中的容器，docker ps -a -q 命令将返回所有的容器 ID
$ docker rm -f $(docker ps -a -q)
```

（3）如果想导入/导出容器，可以使用 docker import 或者 docker export 命令：

```
### 导出容器为 tar 文件，若不指定导出的 tar 路径，默认为当前目录
$ docker export 913111e2d596 > centos.tar
```

导出的 centos 容器包可随时在另一台 Docker 机器上导入为镜像，具体命令如下：

```
$ docker import centos.jar  centos:latest
```

18.3 制作与自动化构建镜像

18.3.1 制作镜像

上一节我们已经学习了如何使用 Docker 命令来操作镜像和容器，这一节学习如何制作 Java 运行环境的镜像，并在此镜像上启动 Java 容器。具体步骤如下：

步骤 01 下载 JDK 安装包，下载地址为 https://www.oracle.com/technetwork/java/javase/downloads/jdk11-downloads-5066655.html。这里使用的 JDK 版本为：jdk-11.0.2_linux-x64_bin.tar.gz。

步骤 02 拉取 centos 镜像，并启动 centos 容器：

```
### 拉取centos镜像
$ docker pull centos
### 启动centos容器
$ docker run -i -t -v /c/Users/Ay:/mnt/ centos /bin/bash
### 查看路径是否挂载成功
[root@8326ca477b44 /]# ll /mnt/
total 950688
-rw-r--r-- 1 root root 179640645 Feb  1 06:00 jdk-11.0.2_linux-x64_bin.tar.gz

// 省略大量代码
```

- -v选项：-v在Docker中称为数据卷（Data Volume），用于将宿主机上的磁盘挂载到容器中，格式为"宿主机路径:容器路径"，需要注意的是宿主机路径可以是相对路径，但是容器的路径必须是绝对路径。可以多次使用 -v 选项，同时挂载多个宿主机路径到容器中。
- /c/Users/Ay:/mnt/：/c/Users/Ay为宿主机JDK安装包的存放路径，/mnt/为centos容器的目录。

步骤 03 在 centos 容器的/mnt/目录下解压安装包并安装 JDK：

```
### 将压缩包解压到 /opt 目录下
[root@8326ca477b44 /]# tar -zxf /mnt/jdk-11.0.2_linux-x64_bin.tar.gz -C /opt
配置环境变量
### 设置环境变量，进入profile文件
[root@8326ca477b44 bin]# vi /etc/profile
```

在 profile 文件末尾添加如下配置：

```
### JAVA_HOME是java的安装路径
JAVA_HOME=/opt/jdk-11.0.2
PATH=$JAVA_HOME/bin:$PATH:.
CLASSPATH=$JAVA_HOME/lib/tools.jar:$JAVA_HOME/lib/dt.jar:.
```

```
export JAVA_HOME
export PATH
export CLASSPATH
```

执行 source 命令,让配置生效:

```
[root@8326ca477b44 bin]# source /etc/profile
```

最后验证 JDK 是否安装成功:

```
### 查看JDK是否安装成功,从输出的信息可知JDK安装成功
[root@8326ca477b44 /]# java -version
java version "11.0.2" 2019-01-15 LTS
Java(TM) SE Runtime Environment 18.9 (build 11.0.2+9-LTS)
Java HotSpot(TM) 64-Bit Server VM 18.9 (build 11.0.2+9-LTS, mixed mode)
```

步骤 04 再打开一个命令行终端,通过 docker commit 命令提交当前容器为新的镜像:

```
### 查看当前运行的容器
$ docker ps
CONTAINER ID    IMAGE     COMMAND       CREATED        STATUS        PORTS       NAMES
8326ca477b44    centos    "/bin/bash"   2 hours ago    Up 2 hours                infallible_kare
### 提交当前容器为新的镜像
$ docker commit 8326ca477b44 hwy/centos
sha256:324e55254ad9baa74477c08333bed2978e72051b7d3f77c957b773d25cfbe7c7
### 查看当前镜像,可知镜像已经成功生成
$ docker images
REPOSITORY      TAG        IMAGE ID           CREATED           SIZE
hwy/centos      latest     324e55254ad9       5 seconds ago     504MB
```

步骤 05 验证生成的镜像是否可用。

```
$ docker run --rm hwy/centos /opt/jdk-11.0.2/bin/java -version
java version "11.0.2" 2019-01-15 LTS
Java(TM) SE Runtime Environment 18.9 (build 11.0.2+9-LTS)
Java HotSpot(TM) 64-Bit Server VM 18.9 (build 11.0.2+9-LTS, mixed mode)
```

上述命令中,我们在 hwy/centos 镜像上启动一个容器,并在容器目录/opt/jdk-11.0.2/bin/下执行命令 java -version,可以看到命令行终端输出了 JDK 版本号相关信息。需要注意的是,上述命令添加了一个 --rm 选项,该选项表示容器退出时可自动删除容器。

至此,手工制作 Docker 镜像已完成。

18.3.2　使用 Dockerfile 构建镜像

从上一节的 Docker 命令学习中可以了解到，镜像的定制实际上就是定制每一层所添加的配置和文件。如果可以把每一层修改、安装、构建、操作的命令都写入一个脚本，用这个脚本来构建、定制镜像，实现整个过程的自动化，既可以提高效率，又能够减少错误。这个脚本就是 Dockerfile。

Dockerfile 是一个文本文件，其内包含了一条条的指令（Instruction），每一条指令构建一层，因此每一条指令的内容，就是描述该层应当如何构建。下面我们创建一个空白文件（文件名为 Dockerfile），并学习 Dockerfile 指令。

1. FROM指令

所谓定制镜像，一定是以一个镜像为基础，在其上进行定制。就像我们之前运行了一个 centos 镜像的容器，再进行修改一样，基础镜像是必须指定的，而 FROM 命令就是用于指定基础镜像。因此一个 Dockerfile 中 FROM 是必备的指令，并且必须是第一条指令。例如：

```
FROM centos:latest
```

FROM 命令的值有固定的格式：即"仓库名称：标签名"，若使用基础镜像的最新版本，则 latest 标签名可以省略，否则需指定基础镜像的具体版本。

2. MAINTAINER指令

MAINTAINER 用于设置该镜像的作者，具体格式：MAINTAINER <author name>。例如：

```
### 建议使用 姓名+邮箱 的形式
MAINTAINER "hwy"<huangwenyi10@163.com>
```

3. ADD指令

ADD 是复制文件指令，它有两个参数 <source> 和 <destination>。source 参数为宿主机的来源路径，destination 是容器内的路径，必须为绝对路径。语法为 ADD <src> <destination>。例如：

```
### 添加jdk安装包到容器的 /opt 目录下
ADD /c/Users/Ay/jdk-11.0.2_linux-x64_bin.tar.gz /opt
```

ADD 指令将自动解压来源中的压缩包，将解压后的文件复制到目标目录（/opt）中。

4. RUN指令

RUN 指令用来执行一系列构建镜像所需要的命令。如果需要执行多条命令，可以使用多条 RUN 指令，例如：

```
### 执行shell命令
RUN echo 'hello ay...'
RUN ls -l
...
```

Dockerfile 中每一个指令都会建立一层，RUN 指令也不例外。每一个 RUN 的行为就和我们手工建立镜像的过程一样，即新建立一层，在其上执行这些命令，构成新的镜像。上面的这种写法，创建了多层镜像，但这是完全没有意义的，其结果就是产生非常臃肿、非常多层的镜像，不仅增加了构建部署的时间，也很容易出错。这是很多 Docker 初学者常犯的一个错误。

5. CMD指令

CMD 指令提供了容器默认的执行命令。Dockerfile 只允许使用一次 CMD 指令，使用多个 CMD 指令会抵消之前所有的指令，只有最后一个指令生效。CMD 指令有以下 3 种形式：

```
CMD ["executable","param1","param2"]
CMD ["param1","param2"]
CMD command param1 param2
```

例如，使用如下指令在容器启动时输出 Java 版本：

```
### 容器启动时执行的命令
CMD /opt/jdk-11.0.2/bin/java -version
```

熟悉了 Dockerfile 文件指令后，下面我们使用 Dockerfile 构建一个 Java 镜像。具体步骤如下：

步骤 01 在用户目录下（备注：C:\Users\Ay）创建 dockerfile 文件夹，在 dockerfile 文件夹中创建并编辑 Dockerfile 文件，同时把之前下载的 jdk-11.0.2_linux-x64_bin.tar.gz 文件放入 dockerfile 文件夹中。

步骤 02 在 Dockerfile 文件中添加如下命令，这些命令都是之前学习 Dockerfile 指令用到的。

```
FROM centos:latest
MAINTAINER "hwy"<huangwenyi10@163.com>
```

```
ADD jdk-11.0.2_linux-x64_bin.tar.gz /opt
RUN echo 'hello ay...'
CMD /opt/jdk-11.0.2/bin/java -version
```

 将 Dockerfile 文件与需要添加到容器的文件放在同一个目录下。

步骤 03 使用 docker bulid 命令读取 Dockerfile 文件，并构建镜像：

```
C:\Users\Ay\dockerfile>docker build -t hwy/java -f Dockerfile .
Sending build context to Docker daemon  179.6MB
Step 1/5 : FROM centos:latest
 ---> 9f38484d220f
Step 2/5 : MAINTAINER "hwy"<huangwenyi10@163.com>
 ---> Running in ca15590e3730
Removing intermediate container ca15590e3730
 ---> 74ae85fd1d25
Step 3/5 : ADD jdk-11.0.2_linux-x64_bin.tar.gz /opt
 ---> 4ab75fb4cd0f
Step 4/5 : RUN echo 'hello ay...'
 ---> Running in 61e64cc0d845
hello ay...
Removing intermediate container 61e64cc0d845
 ---> 3ea28a5e6b05
Step 5/5 : CMD /opt/jdk-11.0.2/bin/java -version
 ---> Running in 35d1544278ed
Removing intermediate container 35d1544278ed
 ---> 51e1a6888f9f
Successfully built 51e1a6888f9f
Successfully tagged hwy/java:latest
SECURITY WARNING: You are building a Docker image from Windows against a
non-Windows Docker host. All files and directories added to build context
will have '-rwxr-xr-x' permissions. It is recommended to double check and
reset permissions for sensitive files and directories.
```

- -t 选项：用于指定镜像的名称，并读取当前目录（即 . 目录）中的 Dockerfile 文件。
- -f 选项：用于指定 Dockerfile 文件名称。

从输出信息可知，执行 docker build 命令后，首先构建上下文发送到 Docker 引擎中，随后通过 5 个步骤来完成构建镜像的构建工作，在每个步骤中都会输出对应的 Dockerfile 命令，而且每个步骤都会生成一个"中间容器"与"中间镜像"。例如步骤 5：

```
Step 5/5 : CMD /opt/jdk-11.0.2/bin/java -version
### 生成中间容器
 ---> Running in 35d1544278ed
### 删除中间容器
Removing intermediate container 35d1544278ed
### 创建一个中间镜像
 ---> 51e1a6888f9f
```

当执行完命令 CMD /opt/jdk-11.0.2/bin/java -version 后，将生成一个中间容器，容器 ID 为 9d10dbefcf6c，接着从该容器中创建一个中间镜像，镜像 ID 为 35d1544278ed，最后将中间容器删除。

> 并不是每个步骤都会生成中间容器，但是每个步骤一定会产生中间镜像。这些中间镜像将加入到缓存中，当某一个构建步骤失败时，将停止整个构建过程，但是中间镜像仍然会存放在缓存中，下次再次构建时，直接从缓存中获取中间镜像，而不会重复执行之前已经构建成功的步骤。

步骤 04 查看生成的 Docker 镜像。

```
### 查看生成的 Docker 镜像
C:\Users\Ay\dockerfile>docker images
REPOSITORY       TAG          IMAGE ID          CREATED           SIZE
hwy/java         latest       51e1a6888f9f      3 minutes ago     504MB
hwy/centos       latest       39969e9d9569      27 hours ago      504MB
tomcat           latest       238e6d7313e3      4 days ago        506MB

centos           latest       9f38484d220f      4 months ago      202MB
hello-world      latest       fce289e99eb9      6 months ago      1.84kB
```

步骤 05 至此，我们完成了通过 Dockerfile 构建镜像。

18.4 Spring Boot 集成 Docker

本节我们开始学习如何在 Spring Boot 中集成 Docker，并在构建 Spring Boot 应用程序时生成 Docker 镜像。具体步骤如下：

步骤 01 创建一个 Spring Boot 项目，项目名为 spring-boot-docker，具体步骤参考 1.3 节。打开 pom.xml 文件，修改 artifactId 和 version，具体代码如下：

```xml
<?xml version="1.0" encoding="UTF-8"?>
<project xmlns="http://maven.apache.org/POM/4.0.0"
xmlns:xsi="http://www.w3.org/2001/XMLSchema-instance"
    xsi:schemaLocation="http://maven.apache.org/POM/4.0.0 http://maven.apache.org/xsd/maven-4.0.0.xsd">
    <modelVersion>4.0.0</modelVersion>
    <parent>
        <groupId>org.springframework.boot</groupId>
        <artifactId>spring-boot-starter-parent</artifactId>
        <version>2.1.6.RELEASE</version>
        <relativePath/> <!-- lookup parent from repository -->
    </parent>
    <groupId>com.example</groupId>
    <!-- 修改 artifactId -->
    <artifactId>spring-boot-docker</artifactId>
    <!-- 修改 version -->
    <version>0.0.1</version>
    <name>demo</name>
    <description>Demo project for Spring Boot</description>

    <properties>
        <java.version>1.8</java.version>
    </properties>

    <dependencies>
        <dependency>
            <groupId>org.springframework.boot</groupId>
```

```xml
        <artifactId>spring-boot-starter-web</artifactId>
    </dependency>

    <dependency>
        <groupId>org.springframework.boot</groupId>
        <artifactId>spring-boot-starter-test</artifactId>
        <scope>test</scope>
    </dependency>
</dependencies>

<build>
    <!-- spring boot maven 插件 -->
    <plugins>
        <plugin>
            <groupId>org.springframework.boot</groupId>
            <artifactId>spring-boot-maven-plugin</artifactId>
        </plugin>
    </plugins>
</build>

</project>
```

Spring Boot 使用 spring-boot-maven-plugin 插件构建项目,通过使用 mvn package 命令打包后将生成一个可直接运行的 jar 包,jar 包默认文件名格式为 ${project.build.finalName},这是一个 Maven 属性,相当于 ${project.artifacId}-${project.version}.jar,生成的 jar 包在/target 目录下。根据上面 pom 文件的配置,执行 mvn package 命令后,生成的 jar 包名为 spring-boot-docker-0.01.jar。

步骤02 在 pom.xml 文件中加入 docker-maven-plugin 插件依赖,具体代码如下:

```xml
<properties>
    <java.version>1.8</java.version>
    <docker.image.prefix>springboot</docker.image.prefix>
</properties>
<build>
    <plugins>
        <plugin>
            <groupId>org.springframework.boot</groupId>
            <artifactId>spring-boot-maven-plugin</artifactId>
```

```xml
        </plugin>
        <!-- Docker maven plugin -->
        <plugin>
            <groupId>com.spotify</groupId>
            <artifactId>docker-maven-plugin</artifactId>
            <version>1.0.0</version>
            <configuration>
                <!-- 指定 Docker 镜像完整名称 -->
                <imageName>${docker.image.prefix}/${project.artifactId}</imageName>
                <!-- 指定 dockerfile 文件所在目录 -->
                <dockerDirectory>src/main/docker</dockerDirectory>
                <resources>
                    <resource>
                        <targetPath>/</targetPath>
                        <directory>${project.build.directory}</directory>
                        <include>${project.build.finalName}.jar</include>
                    </resource>
                </resources>
            </configuration>
        </plugin>
        <!-- Docker maven plugin -->
    </plugins>
</build>
```

- <imageName>：用于指定 Docker 镜像的完整名称。其中${docker.image.prefix}为仓库名称，${project.artifactId}为镜像名。
- <dockerDirectory>：用于指定 dockerfile 文件所在目录。
- <directory>：用于指定需要复制的根目录，其中${project.build.directory}表示target 目录。
- <include>：用于指定需要复制的文件，即 Maven 打包后生成的 jar 文件。

步骤 03 在 src/main/docker/目录下创建 Dockerfile 文件，Dockerfile 文件内容如下：

```
### 使用Docker提供的Java镜像
FROM java
### 作者信息：用户名 + 邮箱
MAINTAINER "hwy"huangwenyi10@163.com
### 复制文件并且重命名为app.jar
ADD spring-boot-docker-0.0.1.jar app.jar
### 将8080端口设置为可暴露的接口
EXPOSE 8080
### 使用java -jar启动项目
CMD java -jar app.jar
```

使用mvn docker:build命令构建项目

```
### 在项目spring-boot-docker目录下执行命令
➜ spring-boot-docker >> mvn docker:build
```

命令执行后，可在控制台上看到相关的输出信息。执行 docker images 命令查看镜像是否成功生成。

```
D:\MyJob\spring-boot-docker>docker images
REPOSITORY                    TAG       IMAGE ID       CREATED          SIZE
springboot/spring-boot-docker latest    b705802f2dd4   48 seconds ago   660MB
hwy/java                      latest    51e1a6888f9f   23 hours ago     504MB
hwy/centos                    latest    39969e9d9569   2 days ago       504MB
tomcat                        latest    238e6d7313e3   5 days ago       506MB
centos                        latest    9f38484d220f   4 months ago     202MB
hello-world                   latest    fce289e99eb9   6 months ago     1.84kB
java                          latest    d23bdf5b1b1b   2 years ago      643MB
```

步骤 04 执行 docker run 命令启动容器。

容器在启动的时候，会执行 Dockerfile 文件里的 CMD 命令：java -jar app.jar，该命令用来启动 Spring Boot 项目，之后就可以在控制台上看到 Spring Boot 的启动信息，具体内容如下：

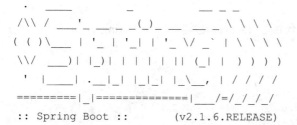

```
  .   ____          _            __ _ _
 /\\ / ___'_ __ _ _(_)_ __  __ _ \ \ \ \
( ( )\___ | '_ | '_| | '_ \/ _` | \ \ \ \
 \\/  ___)| |_)| | | | | || (_| |  ) ) ) )
  '  |____| .__|_| |_|_| |_\__, | / / / /
 =========|_|==============|___/=/_/_/_/
 :: Spring Boot ::        (v2.1.6.RELEASE)
```

```
2019-07-23 14:57:37.624  INFO 5 --- [           main]
com.example.demo.DemoApplication         : Starting DemoApplication v0.0.1
on e7db0dcedf96 with PID 5 (/app
.jar started by root in /)
2019-07-23 14:57:37.658  INFO 5 --- [           main]
com.example.demo.DemoApplication         : No active profile set, falling
back to default profiles: default
2019-07-23 14:57:44.868  INFO 5 --- [           main]
o.s.b.w.embedded.tomcat.TomcatWebServer  : Tomcat initialized with port(s):
8080 (http)
2019-07-23 14:57:45.059  INFO 5 --- [           main]
o.apache.catalina.core.StandardService   : Starting service [Tomcat]
2019-07-23 14:57:45.060  INFO 5 --- [           main]
org.apache.catalina.core.StandardEngine  : Starting Servlet engine:
[Apache Tomcat/9.0.21]
2019-07-23 14:57:45.773  INFO 5 --- [           main]
o.a.c.c.C.[Tomcat].[localhost].[/]       : Initializing Spring embedded
WebApplicationContext
2019-07-23 14:57:45.773  INFO 5 --- [           main]
o.s.web.context.ContextLoader            : Root WebApplicationContext:
initialization completed in 7693 ms
2019-07-23 14:57:46.943  INFO 5 --- [           main]
o.s.s.concurrent.ThreadPoolTaskExecutor  : Initializing ExecutorService
'applicationTaskExecutor'
2019-07-23 14:57:47.956  INFO 5 --- [           main]
o.s.b.w.embedded.tomcat.TomcatWebServer  : Tomcat started on port(s): 8080
(http) with context path ''
2019-07-23 14:57:47.970  INFO 5 --- [           main]
com.example.demo.DemoApplication         : Started DemoApplication in 12.77
seconds (JVM running for 15.003
)
```

步骤05 至此，Spring Boot 成功集成了 Docker 容器，并把 Spring Boot 项目打包成 Docker 镜像。

第 19 章 微服务测试

本章主要介绍 Spring Boot 单元测试、Mockito/PowerMockito 测试框架、H2 内存型数据库、REST API 测试以及性能测试等内容。

19.1 Spring Boot 单元测试

19.1.1 关于测试

软件测试的目的是保证程序员编写的程序达到预期的结果，保证发布的产品是产品经理（产品设计人员）的真实意愿表现。这些都需要软件测试来监督实现，避免将有缺陷的软件发布到生产环境。

软件测试的种类很多，粗略地可划分为单元测试、集成测试、端到端测试。从其他的角度来说，又有回归测试、自动化测试、性能测试等。当我们的项目进行服务化改造之后，尤其是进行了微服务设计之后，测试工作会变得更加困难。很多项目都是以独立服务的形式发布的，这些服务的发布如何保证已经进行充分测试？测试的入口应该在哪里？是直接进行集成测试，还是做端到端的用户体验测试？好像都不太合适。按照分层测试的思想，于是就有了服务测试的话题。微服务的测试理论和其他的测试应该是大体类似的，其中比较特殊的是，如何提供方便快捷的服务测试入口。

目前常见的微服务设计都采用分布式服务框架，这些框架从通信协议上可分为两种：

（1）基于公共标准的 HTTP 协议的

第一种以 HTTP 协议的微服务接口比如使用 Spring Boot 开发的服务，这样的服务测试工具有很多，比如 Postman、Swagger 是常用的工具。如果想为测试人员做点事情的话，可以根据服务注册中心做一个所有服务的列表。

（2）基于私有的 RPC 调用协议

第二种是以私有协议暴露的服务测试，相对比较麻烦。为了打通服务接口和测试人员之间的屏障，以便让测试人员方便地测试到 RPC 协议的服务接口，为每个服务接口写一个客户端，将其转换为 HTTP 协议暴露，这是一种解决办法。但是，这样无形中增加了很多工作量，而且测试服务的质量还依赖于客户端编写的质量，明显是费力不讨好的工作。

那么，如何构建一个项目，它能提供所有服务的客户端，这样新开发一个服务只需要做极少的工作就能生成一个服务的测试客户端，从而快速地将接口提交测试，这是我们下面要讨论的问题。

19.1.2　微服务测试

微服务设计的项目一般都是基于分布式服务的注册和发现机制的，所有的服务都是在一个注册中心集中存储的，而且一般的分布式服务框架都支持丰富的服务调用方式，如基于 Spring XML 配置的和 Spring 注解以及 API 等调用方法，为编写公共的服务测试工具提供了便利的条件。

其所设计的服务测试工具在整个分布式服务架构中所扮演的角色如图 19-1 所示。

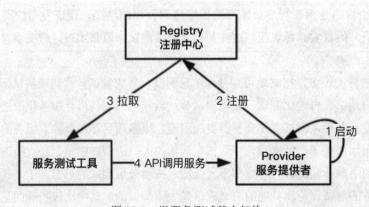

图 19-1　微服务测试基本架构

第 19 章 微服务测试

微服务测试的流程可描述为以下 5 个步骤：

步骤 01 从服务注册中心获取到所有的服务接口，并将这些接口可视化地展示给测试人员，测试人员可以选择需要测试的服务接口，如图 19-2 所示。

序号	项目	服务接口类	接口方法数	服务号	操作
1	商品服务	com.ay.product.ProductService	3	1.0.0	测试
2	商品服务	com.ay.product.PriceService	1	1.0.0	测试
3	商品服务	com.ay.product.ShopService	5	1.0.0	测试
4	用户服务	com.ay.user.UserService	2	1.0.0	测试
5	用户服务	com.ay.user.AddressService	3	1.0.0	测试

图 19-2　服务接口页面展示

步骤 02 将服务的每个接口以一种易读易用的方式暴露给测试人员，比如将接口的请求参数转化为 XML 或者 JSON 的形式展示给测试人员，方便他们输入测试用例。

步骤 03 将测试人员提交的请求参数转换为请求对象，以便使用统一的 API 接口，调用到后端服务。

步骤 04 发起服务调用，使用 API 的方式来调用服务是因为我们做的工具是统一的服务调用入口，能够根据请求参数动态地调用不同的服务。

步骤 05 将服务的响应参数再次转换为 XML 或者 JSON 格式展示给测试人员查看，这时候可以顺便返回调用耗时等附加数据，帮助测试人员判断服务的效率等情况。

可将微服务测试的流程用图 19-3 表示。

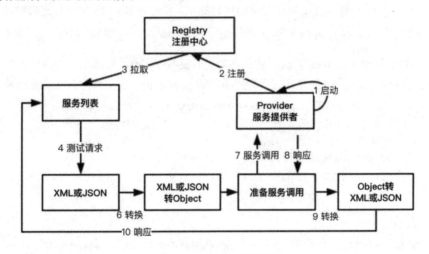

图 19-3　微服务测试流程

微服务测试的宗旨就是尽可能地简化服务测试过程，其中还有一些服务测试基础功能之外的拓展功能：

（1）请求参数的自动化生成，例如请求流水号、请求时间、手机号码、身份证号码等，减少测试人员的输入参数时间。

（2）后台保存服务测试的请求参数和响应参数，方便回归测试。

（3）实现回归测试，在服务代码有变动之后，可根据保存的请求参数进行回归测试，并且可以和之前的响应参数进行对比，以便验证是否影响到当前测试服务接口。

（4）服务的并发测试，在提交测试请求的时候可以指定每个服务测试请求的测试次数，这时后台会模拟多线程调用服务，可实现对服务接口的并发测试。

（5）多个测试环境自由切换，通过选择不同环境的注册中心，来实现其他环境的测试。

（6）服务测试出现异常的时候，将异常堆栈信息直接展示给测试人员，方便排查问题。

（7）实现定时回归测试，有时候我们的测试环境也需要保持一定的稳定性，因为经常会有别的系统发起联调测试。定时回归测试，既能及时发现后端系统对服务的影响，又能保证服务持续稳定地对外提供服务。

（8）开发公共的 mock 测试服务，避免后端未开发完成的服务耽误服务的测试。

19.2　Spring Boot 单元测试

项目在投入生产之前，需要进行大量的单元测试，Spring Boot 作为分布式微服务架构的脚手架，非常有必要来了解下 Spring Boot 如何进行单元测试。具体步骤如下：

步骤 01　创建一个 Spring Boot 项目，项目名为 spring-boot-test，具体步骤参考 1.3 节。

步骤 02　pring-boot-test 项目创建完成后，在项目的 pom.xml 配置文件中，可以看到 Spring Boot 默认已经为我们添加了 spring-boot-starter-test 插件，具体代码如下：

```xml
<dependency>
    <groupId>org.springframework.boot</groupId>
    <artifactId>spring-boot-starter-test</artifactId>
    <scope>test</scope>
</dependency>
```

spring-boot-starter-test 插件依赖了 spring-boot-test、junit、assertj、mockito、hamcrest 等测试框架和类库。

步骤 03　开发用户接口 UserService 和实现类 UserServiceImpl。UserService 接口如下：

```java
/**
 * 描述：用户接口
 * @author ay
 * @date 2019-03-11
 */
public interface UserService {
    AyUser findUser(String id);
}
```

UserServiceImpl 实现类代码如下：

```java
/**
 * 描述：用户服务
 * @author ay
 * @date 2019-03-11
 */
@Component
public class UserServiceImpl implements UserService{
    @Override
    public AyUser findUser(String id) {
        AyUser ayUser = new AyUser();
        ayUser.setId(1);
        ayUser.setName("ay");
        return ayUser;
    }
}
```

用户实体类代码如下：

```java
/**
 * 描述：用户实体类
 * @author ay
 * @date 2019-03-11
 */
public class AyUser {
    private Integer id;
    private String name;
    //省略 set、get 方法
}
```

步骤 04 Spring Boot 的测试类主要放置在/src/test/java 目录下面。项目创建完成后，Spring Boot 会自动生成测试类 DemoApplicationTests.java。测试类的代码如下：

```java
@RunWith(SpringRunner.class)
@SpringBootTest
public class DemoApplicationTests {
    @Resource
    private UserService userService;
    @Test
    public void contextLoads() {}

    @Test
    public void testFindUser(){
        AyUser ayUser = userService.findUser("1");
        Assert.assertNotNull("user is null",ayUser);
    }
}
```

- @RunWith（SpringRunner.class）：@RunWith(Parameterized.class) 参数化运行器，配合 @Parameters 使用 JUnit 的参数化功能。查源代码可知，SpringRunner 类继承 SpringJUnit4ClassRunner 类，此处表明使用 SpringJUnit4ClassRunner 执行器，此执行器集成了 Spring 的一些功能。如果只是简单地 JUnit 单元测试，该注解可以去掉。
- @SpringBootTest：此注解能够测试 SpringApplication，因为 Spring Boot 程序的入口是 SpringApplication，基本上所有配置都会通过入口类去加载，而该注解可以引用入口类的配置。
- @Test：JUnit 单元测试的注解，注解在方法上表示一个测试方法。

当右键执行 DemoApplicationTests.java 中的 contextLoads 方法时，可以看到控制台打印的信息和执行入口类中的 SpringApplication.run()方法打印的信息是一致的。由此便知，@SpringBootTest 是引入了入口类的配置。

在 DemoApplicationTests.java 类中添加测试用例 testFindUser，并在方法上添加@Test注解，运行测试用例，通过使用 JUnit 框架提供的 Assert.assertXXX()断言方法来验证期望值与实际值是否一致。如果不一致，将打印错误信息"user is null"，这就是单元测试的基本做法。

JUnit 框架提供的 Assert 断言一方面需要提供错误信息，另一方面期望值与实际值到底谁在前谁在后，很容易犯错。好在 Spring Boot 已经考虑到这些因素，它依赖于 AssertJ 类库，弥补了 JUnit 框架在断言方面的不足之处。我们可以轻松地将 JUnit 断言修改为 AssertJ 断言，具体代码如下：

```
@Test
public void testFindUser(){
    boolean success = false;
    int num = 10;
    AyUser ayUser = userService.findUser("1");
    //JUnit 断言
    Assert.assertNotNull("user is null",ayUser);
    //AssertJ 断言
    Assertions.assertThat(ayUser).isNotNull();
    //JUnit 断言
    Assert.assertTrue("result is not true", success);
    //AssertJ 断言
    Assertions.assertThat(success).isTrue();
    //JUnit 断言
    Assert.assertEquals("num is not equal 10", 10, num);
    //AssertJ 断言
    Assertions.assertThat(num).isEqualTo(10);
}
```

19.3 Mockito/PowerMockito 测试框架

19.3.1 Mockito 概述

Mockito 是用于生成模拟对象或者直接说就是"假对象"的模拟工具。其特点是对于某些不容易构造（如 HttpServletRequest）或者不容易获取的复杂对象（如 JDBC 中的 ResultSet 对象），可用一个虚拟的对象（Mock 对象）来创建以便完成测试。Mockito 最大的优点是可帮你把单元测试的耦合分解开，如果你的代码对另一个类或者接口有依赖，它能够帮助你模拟这些依赖，并帮助你验证所调用的依赖的行为。我们先来看一个传统的测试用例调用流程图，如图 19-4 所示。

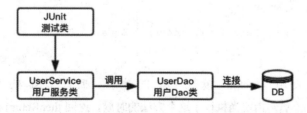

图 19-4 JUnit 传统的测试用例调用流程

从图 19-4 可知，当想要测试用户服务类 UserService 的某些接口时，需要依赖 UserDao 对象来完成相关测试，而 UserDao 对象还需要连接数据库。某些情况下我们无法连接数据库，比如无网络的情况下，此时，测试用例就无法正常执行。清楚了传统 JUnit 测试用例的局限性，我们来看一下 Mockito 如何规避这些缺点，如图 19-5 所示。

图 19-5　Mockito 测试用例调用流程

从图 19-5 可知，利用 Mockito 框架提供的强大模拟对象功能，模拟出 UserDao 对象，并去掉 UserDao 与 DB 连接的关系，可以快速地开发出独立、稳定的测试用例，该测试用例不会因为 DB 异常而导致运行失败。实际中，JUnit 和 Mockito 两者定位不同，项目中通常的做法是联合 JUnit + Mockito 来进行测试。

19.3.2　Mockito 简单实例

上一节，简单了解了 Mockito 的概念和优点，这里列举几个简单实例来体验一下 Mockito。

实例一：

```
@Test
public void testMockito_1() {
    List mock = mock(List.class);
    when(mock.get(0)).thenReturn("ay");
    when(mock.get(1)).thenReturn("al");
    //测试通过
    Assertions.assertThat(mock.get(0)).isEqualTo("ay");
    //测试不通过
    Assertions.assertThat(mock.get(1)).isEqualTo("xx");
}
```

上面实例中，使用 Mockito 模拟 List 的对象，拥有 List 的所有方法和属性。when(xxxx).thenReturn(yyyy)指定当执行了这个方法的时候，返回 thenReturn 的值，相当于是对模拟对象的配置过程，为某些条件给定一个预期的返回值。Mockito 通过 when(xxx).thenReturn(yyy)

这样的语法来定义对象方法和参数（输入），然后在 thenReturn 中指定结果（输出），此过程称为 stub 打桩。一旦这个方法被 stub 了，就会一直返回这个 stub 的值。

stub 打桩时，需要注意以下几点：

（1）对于 static 和 final 方法，Mockito 无法对其 when(…).thenReturn(…)操作。

（2）当连续两次为同一个方法使用 stub 的时候，它只会使用最新的一次。

实例二：

首先，我们开发 AyUser 实体、UserDao 和 UserService 接口、UserServiceImpl 实现类，具体代码如下：

```java
/**
 * 描述: 用户实体类
 * @author ay
 * @date 2019-03-11
 */
public class AyUser {
    private Integer id;
    private String name;

    public AyUser(Integer id, String name) {
        this.id = id;
        this.name = name;
    }

    public AyUser(){}

    //省略 set、get 代码
}

/**
 * 描述: UserDao
 * @author ay
 * @date 2019-03-13
 */
@Component
public class UserDao {
```

```java
    public AyUser findUser(Integer userId){
        AyUser user = null;//查询数据库
        return user;
    }

    public boolean deleteUser(Integer userId){
        //操作数据库
        return true;
    }
}

/**
 * 描述：用户接口
 * @author ay
 * @date 2019-03-11
 */
public interface UserService {
    //查询用户
    AyUser findUser(Integer id);
    //删除用户
    boolean deleteUser(Integer id);
}

/**
 * 描述：用户服务
 * @author ay
 * @date 2019-03-11
 */
@Component
public class UserServiceImpl implements UserService{

    @Resource
    private UserDao userDao;

    @Override
    public AyUser findUser(Integer id) {
        AyUser ayUser = userDao.findUser(id);
        return ayUser;
```

```
        }

        @Override
        public boolean deleteUser(Integer id) {
            boolean isSuccess = userDao.deleteUser(id);
            return isSuccess;
        }
    }
```

然后,我们开发测试用例,具体代码如下:

```
@Test
public void testMockito_2(){
    UserService userService = mock(UserServiceImpl.class);
    when(userService.findUser(1)).thenReturn(new AyUser(1, "ay"));
    //通过mock,查询出模拟用户对象
    AyUser ayUser = userService.findUser(1);
    //删除用户
    boolean isSuccess = userService.deleteUser(ayUser.getId());

    Assertions.assertThat(isSuccess).isFalse();
}
```

在 testMockito_2 测试用例方法中,当 mock 对象 UserServiceImpl 查询用户的时候返回 mock 对象 new AyUser(1,"ay"),最后删除用户对象。

本节列举的实例非常简单,更多 Mockito 资料请参考官方文档(https://static.javadoc.io/org.mockito/mockito-core/2.25.0/org/mockito/Mockito.html)。读者可根据官方文档,编写出适合自己业务需求的测试用例,在之后的工作中,可以使用该测试框架模拟依赖,简化单元测试中复杂的依赖关系。

19.3.3 PowerMock 概述

Mockito 由于其可以极大地简化单元测试的书写过程而被许多人应用在自己的工作中,但是 Mockito 工具不可以实现对静态函数、构造函数、私有函数、Final 函数以及系统函数的模拟,但是这些方法往往是我们在大型系统中需要的功能。

PowerMock 就是在 Mockito 基础上扩展而来,通过定制类加载器等技术,PowerMock 实现了上述所有模拟功能,使其成为分布式微服务架构必备的单元测试工具。

19.3.4 PowerMockito 简单实例

PowerMock 有两个重要的注解：

（1）@RunWith(PowerMockRunner.class)

（2）@PrepareForTest({ YourClassWithEgStaticMethod.class })

如果测试用例里没有使用注解 @PrepareForTest，那么可以不用加注解 @RunWith(PowerMockRunner.class)，反之亦然。当需要使用 PowerMock 的强大功能（Mock 静态、final、私有方法等）的时候，就需要加注解 @PrepareForTest。使用 PowerMock 之前，需要在项目的 pom.xml 文件中添加依赖信息，具体代码如下：

```
<properties>
        <org.powermock.version>1.7.0</org.powermock.version>
</properties>

<dependency>
        <groupId>org.powermock</groupId>
        <artifactId>powermock-api-mockito</artifactId>
        <scope>test</scope>
        <version>${org.powermock.version}</version>
</dependency>
<dependency>
        <groupId>org.powermock</groupId>
        <artifactId>powermock-module-junit4</artifactId>
        <scope>test</scope>
        <version>${org.powermock.version}</version>
</dependency>
```

接下来，我们来看具体的实例：

```
/**
 * 描述：PowerMockio
 * @author ay
 * @date 2019-05--2
 */
public class PowerMockioTest {
```

```java
        Logger logger = LoggerFactory.getLogger(PowerMockioTest.class);

    @Test
    public void testFindUser() throws Exception{
        //mock 对象
        UserService userService = PowerMockito.spy(new UserService());

        //设置 MAX_TIME = 100
        Whitebox.setInternalState(userService, "MAX_TIME", new AtomicInteger(100));
        String name = "ay";
        //模拟调用 getUserFromDB 方法，返回 new User(1, "ay")对象
        PowerMockito.when(userService.getUserFromDB()).thenReturn(new User(1, "ay"));
        Assert.assertEquals(userService.findUser("ay").getName(), "ay");
        //设置 MAX_TIME = 130
        Whitebox.setInternalState(userService, "MAX_TIME", new AtomicInteger(130));
        try{
            //调用 findUser 方法
            PowerMockito.when(userService, "findUser", name);
        }catch (Exception e){
            logger.error(e.getMessage());
        }

    }
}

/**
 * 描述：用户服务
 * @author ay
 * @date 2019-05-01
 */
class UserService{

    //日志
    Logger logger = LoggerFactory.getLogger(UserService.class);
```

```java
//当前调用次数
public AtomicInteger MAX_TIME;

public User findUser(String name) throws Exception{
    //findUser方法一天只能调用120次
    if(MAX_TIME.get() > 120){
        throw new Exception("系统繁忙");
    }
    //模拟从数据库中查询到的数据
    User user = getUserFromDB();
    Integer maxTime = MAX_TIME.getAndIncrement();
    //记录日志
    logger.info("the current time is :" + maxTime);
    return user;
}

public AtomicInteger getMAX_TIME() {
    return MAX_TIME;
}

public void setMAX_TIME(AtomicInteger MAX_TIME) {
    this.MAX_TIME = MAX_TIME;
}

public User getUserFromDB(){
    return new User(1, "al");
}
}

class User{

    private Integer id;
    private String name;

    public Integer getId() {
        return id;
    }
```

```java
    public void setId(Integer id) {
        this.id = id;
    }

    public String getName() {
        return name;
    }

    public void setName(String name) {
        this.name = name;
    }

    public User(Integer id, String name) {
        this.id = id;
        this.name = name;
    }
}
```

上述实例中，PowerMockito.spy 用来模拟对象，Whitebox.setInternalState 用来模拟给对象设置值，PowerMockito.when 用来模拟方法内部的逻辑。

19.4　H2 内存型数据库

19.4.1　H2 概述

H2 是一个开源的、内存型嵌入式（非嵌入式设备）数据库引擎，它是一个用 Java 开发的类库，可直接嵌入到应用程序中，与应用程序一起打包发布，不受平台限制。更多 H2 的资料请参考官方文档（http://www.h2database.com/html/tutorial.html）。

19.4.2　Spring Boot 集成 H2

步骤01 创建一个 Spring Boot 项目，项目名为 spring-boot-h2，具体步骤参考 1.3 节。

步骤02 在 spring-boot-h2 项目的 pom.xml 文件中添加 H2 的依赖，具体代码如下：

```xml
<dependency>
    <groupId>com.h2database</groupId>
```

```xml
    <artifactId>h2</artifactId>
    <scope>runtime</scope>
</dependency>
<dependency>
    <groupId>org.springframework.boot</groupId>
    <artifactId>spring-boot-starter-data-jpa</artifactId>
</dependency>
<dependency>
    <groupId>org.projectlombok</groupId>
    <artifactId>lombok</artifactId>
</dependency>
```

- spring-boot-starter-data-jpa 依赖：Spring Data JPA 是 Spring Data 的一个子项目，它通过提供基于 JPA 的 Respository，极大地减少了 JPA 作为数据访问方案的代码量。通过 Spring Data JPA 框架，开发者可以省略实现持久层业务逻辑的工作，唯一要做的就只是声明持久层的接口，其他都交给 Spring Data JPA 来帮你完成。
- lombok 依赖：Lombok 能以简单的注解形式来简化 Java 代码，提高开发人员的开发效率。例如开发中经常需要写 JavaBean，都需要花时间去添加相应的 getter/setter 方法，也许还要去写构造器、equals 等方法。这些显得很冗长也没有太多技术含量，一旦修改属性，就容易出现忘记修改对应方法的失误。

Lombok 能通过注解的方式，在编译时自动为属性生成构造器、getter/setter、equals、hashcode、toString 等方法。在源代码中没有 getter 和 setter 方法，但是在编译生成的字节码文件中有 getter 和 setter 方法。这样就省去了手动重建这些代码的麻烦，使代码看起来更简洁。

步骤 03 在/resources 目录下创建配置文件 application-test.properties，并添加如下配置：

```
### 是否生成ddl语句
spring.jpa.generate-ddl=false
### 是否打印sql语句
spring.jpa.show-sql=true
### 自动生成ddl，由于指定了具体的ddl，此处设置为none
spring.jpa.hibernate.ddl-auto=none

### 使用H2数据库
spring.datasource.platform=h2
### H2驱动
spring.datasource.driverClassName =org.h2.Driver
```

```
### 指定生成数据库的 schema 文件位置
spring.datasource.schema=classpath:/db/schema.sql
### 指定插入数据库语句的脚本位置
spring.datasource.data=classpath:/db/data.sql
```

步骤 04 在 resources/db 目录下创建 data.sql 文件和 schema.sql 文件，schema.sql 用于定义数据库表的结构，data.sql 为数据库表的初始化数据。

schema.sql 文件内容如下：

```sql
CREATE TABLE `ay_user` (
    `id` bigint(11) unsigned NOT NULL AUTO_INCREMENT,
    `name` varchar(11) DEFAULT NULL,
    `url` varchar(200) DEFAULT NULL,
    PRIMARY KEY (`id`)
) ENGINE=InnoDB DEFAULT CHARSET=utf8;
```

data.sql 文件内容如下：

```sql
INSERT INTO ay_user (id, name,url) VALUES (1, 'ay', 'https://huangwenyi.com');
INSERT INTO ay_user (id, name,url) VALUES (2, 'al','https://al.com');
```

上述代码中，我们创建了用户表 ay_user，同时往表里插入 2 条数据。随着项目启动，数据初始化到内存中，停止项目，数据消失。

步骤 05 开发 UserRepository 和 User 类，具体代码如下：

```java
/**
 * 描述: UserRepository
 * @author ay
 * @date 2019-03-13
 */
@Repository
public interface UserRepository extends JpaRepository<User, Long> {
    User findByName(String name);
}

/**
 * 描述: 用户实体
 * @author ay
 * @date 2019-03-13
```

```
 */
@Entity
@Table(name = "ay_user")
@Data
public class User {
    @Id
    @GeneratedValue(strategy = GenerationType.IDENTITY)
    private Long id;
    private String name;
    private String url;
}
```

上述代码中,我们创建了 ay_user 表对应的实体类 User,同时开发了 UserRepository 类,用来与 H2 数据库交互,查询数据。类中定义了 findByName 方法,作用是通过用户名查询用户。

在测试类中开发测试用例,具体代码如下:

```
@RunWith(SpringRunner.class)
@SpringBootTest
@TestPropertySource("classpath:application-test.properties")
public class DemoApplicationTests {
    @Test
    public void contextLoads() {}
    @Resource
    private UserRepository userRepository;

    @Test
    public void testSave() throws Exception {
        User user = new User();
        user.setName("ay");
        user.setUrl("https://huangwenyi.com");
        User result = userRepository.save(user);
        Assertions.assertThat(result).isNotNull();
    }

    @Test
    public void testFindOne() throws Exception{
        User user = userRepository.findById(1L).get();
```

```java
        Assertions.assertThat(user).isNotNull();
        Assertions.assertThat(user.getId()).isEqualTo(1);
    }

    @Test
    public void testFindByName() throws Exception{
        User user = userRepository.findByName("ay");
        Assertions.assertThat(user).isNotNull();
        Assertions.assertThat(user.getName()).isEqualTo("ay");
    }
}
```

- @TestPropertySource：该注解可以用来指定读取的配置文件，目前该测试类读取的配置文件为 application-test.properties。
- testSave 方法：测试用户保存是否成功。
- testFindOne 方法：测试通过用户 id 查询用户。
- testFindByName 方法：测试通过用户名查询用户。

逐个执行测试用例，查看测试结果。

19.5 REST API 测试

19.5.1 Postman 概述

Postman 是一款功能强大的网页调试和模拟发送 HTTP 请求的 Chrome 插件，支持几乎所有类型的 HTTP 请求，操作简单且方便。

19.5.2 Postman 的简单使用

接下来，我们学习如何通过 Postman 测试 REST API，具体步骤如下：

步骤 01 创建 Spring boot 项目，项目名称为 spring-boot-postman，具体步骤参考 1.3 节。

步骤 02 下载 Postman 软件，下载地址：https://www.getpostman.com/，具体如图 19-6 所示。下载完成后，按照默认安装即可。

图 19-6　postman 下载页面

步骤 03　安装完成后打开软件，界面如图 19-7 所示。

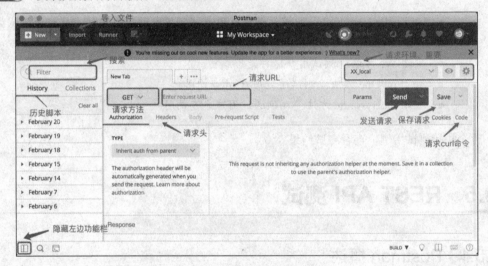

图 19-7　postman 下载地址

- 左侧功能栏：History 为近期发起请求的历史记录，Collections 集合用于管理需要调用的请求集合，也可以新建文件夹，用于放置不同请求的文件集合。
- 主界面：可以选择 HTTP 请求的方法，输入 URL、参数，cookie 管理、脚本保存和另存为等功能。在主界面的右上侧，可以设置不同的环境变量，以满足不同环境的测试需求，这个功能在真实的项目中，频繁被使用。

步骤 04　创建 AyController 控制层类，具体代码如下：

```
/**
 * 描述：控制层
 * @author ay
 * @date 2019-03-17
```

```java
 */
@RestController
@Controller
public class AyController {

    @RequestMapping("/say")
    public String say(Model model){
        return "hello ay";
    }

    @PostMapping("/save")
    public String save(Model model, @RequestBody User user){
        System.out.println(model);
        return "save" + user.name + "success";
    }

    class User{
        private String name;
        //省略 set、get 方法
    }
}
```

步骤 05 使用 Postman 发起 Post 和 Get 请求,如图 19-8 和图 19-9 所示。

图 19-8　Get 请求实例

图 19-9　Post 请求实例

- Authorization：身份验证，主要用来输入用户名、密码以及一些验签字段。
- form-data：对应信息头multipart/form-data，它将表单数据处理为一条消息，以标签为单元用分隔符分开。既可以上传键值对，也可以上传文件；当上传字段是文件时，会有Content-Type来说明文件类型。
- x-www-form-urlencoded：对应信息头application/x-www-from-urlencoded，会将表单内的数据转换为键值对，比如name=ay。
- raw：可以上传任意类型的文本，比如TEXT、JSON、XML等。
- binary：对应信息头Content-Type:application/octet-stream，只能上传二进制文件，且没有键值对，一次只能上传一个文件。

Postman 软件在工作中经常使用，本节只是简单地带读者入门，更多内容请查询官方文档，地址为 https://learning.getpostman.com/docs/postman/launching_postman/installation_and_updates/。

19.6　性能测试

19.6.1　AB 概述

AB 是 Apache 自带的压力测试工具。AB 非常实用，它不仅可以对 Apache 服务器进行网站访问压力测试，也可以对其他类型的服务器进行压力测试。比如 Nginx、Tomcat、IIS 等。

大型互联网项目，用户流量大，基本要求微服务达到三高要求（高性能、高可用和高并发）。因此，我们需要一些测试服务性能的工具来检验微服务的性能，而 Apache 的 AB 工具就是一款性能测试的利器，在大型互联网项目中被广泛使用。

19.6.2　AB 测试

执行命令：ab --help，可以查 ab 命令参数的详细信息。

```
➜ ab --help
ab: wrong number of arguments
Usage: ab [options] [http[s]://]hostname[:port]/path
Options are:
    -n requests     Number of requests to perform
    -c concurrency  Number of multiple requests to make at a time
    -t timelimit    Seconds to max. to spend on benchmarking
                    This implies -n 50000
    -s timeout      Seconds to max. wait for each response
                    Default is 30 seconds
// 省略代码
```

- -n: 执行的请求个数。默认执行一个请求。
- -c: 一次产生的请求个数（并发数）。默认是一次一个。
- -t: 测试所进行的最大秒数。它可以使对服务器的测试限制在一个固定的总时间内，默认没有时间限制。

ab 命令提供的参数很多，一般使用-c 和-n 参数就基本够用了。例如：

```
➜ ab -n 4 -c 2 https://www.baidu.com/
This is ApacheBench, Version 2.3 <$Revision: 1807734 $>
Copyright 1996 Adam Twiss, Zeus Technology Ltd, http://www.zeustech.net/
Licensed to The Apache Software Foundation, http://www.apache.org/
Benchmarking www.baidu.com (be patient).....done
Server Software:        BWS/1.1
Server Hostname:        www.baidu.com
Server Port:            443
SSL/TLS Protocol:       TLSv1.2,ECDHE-RSA-AES128-GCM-SHA256,2048,128
TLS Server Name:        www.baidu.com
```

```
Document Path:          /
Document Length:        227 bytes

Concurrency Level:      2
Time taken for tests:   0.246 seconds
Complete requests:      4
Failed requests:        0
Total transferred:      3572 bytes
HTML transferred:       908 bytes
Requests per second:    16.24 [#/sec] (mean)
Time per request:       123.115 [ms] (mean)
Time per request:       61.558 [ms] (mean, across all concurrent requests)
Transfer rate:          14.17 [Kbytes/sec] received

Connection Times (ms)
              min  mean[+/-sd] median   max
Connect:       78   93   10.0     98    101
Processing:    22   25    4.0     24     31
Waiting:       22   25    4.0     24     31
Total:        101  118   11.9    123    129

Percentage of the requests served within a certain time (ms)
  50%    123
  66%    123
  75%    129
  80%    129
  90%    129
  95%    129
  98%    129
  99%    129
 100%    129 (longest request)
```

从输出的信息可以看出,百度网站首页的吞吐量为 16.24 个/s,平均响应时间是 123.115ms。上述只是一个简单的实例,具体性能测试需要根据业务需求具体分析。

第 20 章

Spring Boot 原理解析

本章主要回顾了 MySpringApplication 入口类上注解和 run 方法的原理，梳理了 Spring Boot 启动执行的流程和简单分析 spring-boot-starter 起步依赖原理，同时介绍真实项目中的跨域问题、Spring Boot 优雅关闭以及如何将普通 Web 项目改造成 Spring Boot 项目等内容。

20.1 回顾入口类

20.1.1 DemoApplication 入口类

首先，我们先来回顾一下项目 spring-boot-book-v2 的入口类 DemoApplication，具体代码如下：

```
@SpringBootApplication
@ServletComponentScan
@ImportResource(locations={"classpath:spring-mvc.xml"})
@EnableAsync
@EnableRetry
public class DemoApplication {

    public static void main(String[] args) {
```

```
        SpringApplication.run(MySpringBootApplication.class, args);
    }
}
```

在入口类 DemoApplication 中，@SpringBootApplication 和 main 方法是 Spring Boot 为我们自动生成的，其他注解都是我们在学习 Spring Boot 整合其他技术添加上去的。接下来就和大家一起看看@SpringBootApplication 和 SpringApplication.run 方法到底为我们做了些什么。

20.1.2 @SpringBootApplication 的原理

@SpringBootApplication 开启了 Spring 的组件扫描和 Spring Boot 自动配置功能。实际上它是一个复合注解，包含 3 个重要的注解@SpringBootConfiguration、@EnableAutoConfiguration、@ComponentScan，其源代码如下：

```
@Target({ElementType.TYPE})
@Retention(RetentionPolicy.RUNTIME)
@Documented
@Inherited
@SpringBootConfiguration
@EnableAutoConfiguration
@ComponentScan
public @interface SpringBootApplication {

    //省略代码

}
```

- @SpringBootConfiguration注解：标明该类使用Spring基于Java的注解，Spring Boot推荐使用基于Java而不是XML的配置，所以本书的实战例子都是基于Java而不是XML的配置。查看@SpringBootConfiguration源代码，可以看到它就是对@Configuration进行简单的"包装"，然后取名为SpringBootConfiguration。@SpringBootConfiguration源代码如下：

```
@Target(ElementType.TYPE)
@Retention(RetentionPolicy.RUNTIME)
@Documented
@Configuration
public @interface SpringBootConfiguration {

}
```

我们对@Configuration注解并不陌生,它就是JavaConfig形式的Spring IoC容器的配置类使用的那个@Configuration。
- @EnableAutoConfiguration注解:该注解可以开启自动配置的功能。@EnableAutoConfiguration的源代码如下:

```
@Target({ElementType.TYPE})
@Retention(RetentionPolicy.RUNTIME)
@Documented
@Inherited
@AutoConfigurationPackage
@Import({EnableAutoConfigurationImportSelector.class})
public @interface EnableAutoConfiguration {
    //省略代码
}
```

从@EnableAutoConfiguration源代码可以看出,其包含@Import注解。我们知道,@Import注解的主要作用就是借助EnableAutoConfigurationImportSelector将Spring Boot应用所有符合条件的@Configuration配置都加载到当前Spring Boot创建并使用的IoC容器中,IoC容器就是我们所说的Spring应用程序上下文ApplicationContext。学习过Spring框架就知道,Spring框架提供了很多@Enable开头的注解定义,比如@EnableScheduling、@EnableCaching等。这些@Enable开头的注解都有一个共同的功能,就是借助@Import的支持,收集和注册特定场景相关的bean定义。
- @ComponentScan注解:启动组件扫描,开发的组件或bean定义能被自动发现并注入到Spring应用程序上下文。比如我们在控制层添加@Controller注解、服务层添加的@Service注解和@Component注解等,这些注解都可以被@ComponentScan注解扫描到。

Spring Boot 早期的版本中,需要在入口类同时添加这3个注解,但从Spring Boot 1.2.0开始,只要在入口类添加@SpringBootApplication注解即可。

20.1.3　SpringApplication 的 run 方法

除了 @SpringBootApplication 注解外,我们发现入口类的一个显眼的地方,就是SpringApplication.run 方法。在 run 方法中,首先创建一个 SpringApplication 对象实例,然后调用 SpringApplication 的 run 方法。SpringApplication.run 方法的源代码如下:

```java
public class SpringApplication{
    //省略代码
    public ConfigurableApplicationContext run(String... args) {
        StopWatch stopWatch = new StopWatch();
        stopWatch.start();
        ConfigurableApplicationContext context = null;
        FailureAnalyzers analyzers = null;
        configureHeadlessProperty();
            //开启监听器
        SpringApplicationRunListeners listeners = getRunListeners(args);
        listeners.starting();
        try {
            ApplicationArguments applicationArguments =
                            new DefaultApplicationArguments(args);
            ConfigurableEnvironment environment = prepareEnvironment(listeners,
                    applicationArguments);
            Banner printedBanner = printBanner(environment);
                //创建应用上下文
            context = createApplicationContext();
            analyzers = new FailureAnalyzers(context);
                //准备上下文
            prepareContext(context, environment, listeners,
                    applicationArguments, printedBanner);
                //刷新应用上下文
            refreshContext(context);
                //刷新后操作
            afterRefresh(context, applicationArguments);
            listeners.finished(context, null);
            stopWatch.stop();
            if (this.logStartupInfo) {
                new StartupInfoLogger(this.mainApplicationClass)
                    .logStarted(getApplicationLog(), stopWatch);
            }
            return context;
        }
        catch (Throwable ex) {
            handleRunFailure(context, listeners, analyzers, ex);
```

```
            throw new IllegalStateException(ex);
        }
    }
}
```

从源代码可以看出，Spring Boot 首先开启了一个 SpringApplicationRunListeners 监听器，然后通过 createApplicationContext、prepareContext 和 refreshContext 方法创建、准备、刷新应用上下文 ConfigurableApplicationContext，通过上下文加载应用所需的类和各种环境配置等，最后启动一个应用实例。

20.1.4　SpringApplicationRunListeners 监听器

SpringApplicationRunListener 接口规定了 Spring Boot 的生命周期，在各个生命周期广播相应的事件（ApplicationEvent），调用实际的是 ApplicationListener 类。SpringApplicationRunListener 源代码如下：

```
public interface SpringApplicationRunListener {
    //执行 run 方法时触发
    void starting();
    //环境建立好时候触发
    void environmentPrepared(ConfigurableEnvironment environment);
    //上下文建立好的时候触发
    void contextPrepared(ConfigurableApplicationContext context);
    //上下文载入配置时候触发
    void contextLoaded(ConfigurableApplicationContext context);
    //上下文刷新完成后，run 方法执行完之前触发
    void finished(ConfigurableApplicationContext context, Throwable exception);
}
```

ApplicationListener 是 Spring 框架对 Java 中实现的监听器模式的一种框架实现。具体源代码如下：

```
public interface ApplicationListener<E extends ApplicationEvent>
        extends EventListener {
    void onApplicationEvent(E var1);
}
```

ApplicationListener 接口只有一个方法 onApplicationEvent，所以自己的类在实现该接口的时候要实现该方法。如果在上下文 ApplicationContext 中部署一个实现了 ApplicationListener 接口的监听器，每当 ApplicationEvent 事件发布到 ApplicationContext 时，该监听器会得到通知。如果要为 Spring Boot 应用添加自定义的 ApplicationListener，可以通过 SpringApplication.addListeners() 或者 SpringApplication.setListeners() 方法添加一个或者多个自定义的 ApplicationListener。

20.1.5 ApplicationContextInitializer 接口

在 Spring Boot 准备上下文 prepareContext 的时候，会对 ConfigurableApplicationContext 实例做进一步的设置或者处理。prepareContext 的源代码如下：

```
private void prepareContext(ConfigurableApplicationContext context,
        ConfigurableEnvironment environment, SpringApplicationRunListeners listeners,
        ApplicationArguments applicationArguments, Banner printedBanner) {
    context.setEnvironment(environment);
    postProcessApplicationContext(context);
     //对上下文进行设置和处理
    applyInitializers(context);
    listeners.contextPrepared(context);
    if (this.logStartupInfo) {
        logStartupInfo(context.getParent() == null);
        logStartupProfileInfo(context);
    }


    // Add boot specific singleton beans
    context.getBeanFactory().registerSingleton("springApplicationArguments",
            applicationArguments);
    if (printedBanner != null) {
        context.getBeanFactory().
registerSingleton("springBootBanner", printedBanner);
    }

    // Load the sources
    Set<Object> sources = getSources();
    Assert.notEmpty(sources, "Sources must not be empty");
```

```
        load(context, sources.toArray(new Object[sources.size()]));
        listeners.contextLoaded(context);
    }
```

在准备上下文 prepareContext 方法中，通过 applyInitializers 方法对 context 上下文进行设置和处理。applyInitializers 的源代码如下：

```
    protected void applyInitializers(ConfigurableApplicationContext context) {
        for (ApplicationContextInitializer initializer : getInitializers()) {
            Class<?> requiredType = GenericTypeResolver.resolveTypeArgument(
                    initializer.getClass(), ApplicationContextInitializer.class);
            Assert.isInstanceOf(requiredType, context, "Unable to call initializer.");
            initializer.initialize(context);
        }
    }
```

在 applyInitializers 方法中，主要是调用 ApplicationContextInitializer 类的 initialize 方法对应用上下文进行设置和处理。ApplicationContextInitializer 本质上是一个回调接口，用于在 ConfigurableApplicationContext 执行 refresh 操作之前对它进行一些初始化操作。一般情况下，我们基本不需要自定义一个 ApplicationContextInitializer，如果真需要自定义一个 ApplicationContextInitializer，那么可以通过 SpringApplication.addInitializers()设置即可。

20.1.6　ApplicationRunner 与 CommandLineRunner

ApplicationRunner 与 CommandLineRunner 接口执行点是在容器启动成功后的最后一步回调，我们可以在回调方法 run 中执行相关逻辑。ApplicationRunner 的源代码如下：

```
public interface ApplicationRunner {
    void run(ApplicationArguments args) throws Exception;
}
```

CommandLineRunner 的源代码如下：

```
public interface CommandLineRunner {
    void run(String... args) throws Exception;
}
```

在 ApplicationRunner 或 CommandLineRunner 类中，只有一个 run 方法，但是它们的入参不一样，分别是 ApplicationArguments 和可变 String 数组。

如果有多个 ApplicationRunner 或 CommandLineRunner 实现类，而我们需要按一定顺序执行它们的话，可以在实现类上加上 @Order(value=整数值) 注解，Spring Boot 会按照 @Order 中的 value 值从小到大依次执行。

如果想在 Spring Boot 启动的时候运行一些特定的代码，你可以实现接口 ApplicationRunner 或者 CommandLineRunner，这两个接口实现方式一样，它们都只提供了一个 run 方法。

例如：

```
/**
 * @author Ay
 * @create 2019/09/08
 **/
public class MyCommandRunner implements CommandLineRunner {

    @Override
    public void run(String... args) throws Exception {
        //do something
    }
}
```

或者：

```
@Bean
public CommandLineRunner init(){
    return (String ... strings)->{

    };
}
```

20.2 SpringApplication 执行流程

上一节，我们对 SpringApplication 的 run 方法进行了简单的学习，本节再简单地总结一下 Spring Boot 启动的完整流程，具体流程如图 20-1 所示。

第 20 章 Spring Boot 原理解析 | 249

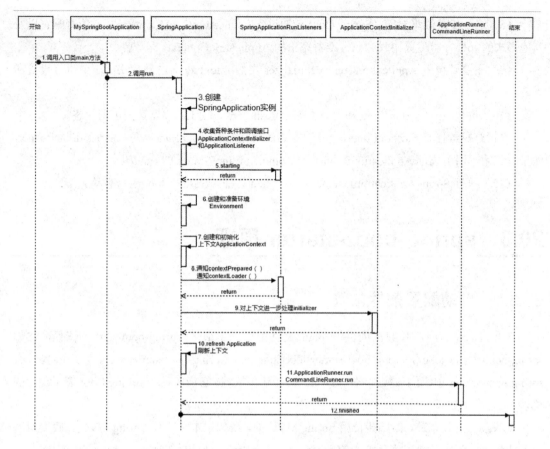

图 20-1　Spring Boot 的启动流程

（1）项目启动时，调用入口类 MySpringBootApplication 的 main 方法。

（2）入口类 MySpringBootApplication 的 main 方法会调用 SpringApplication 的静态方法 run。

（3）在 run 方法中首先创建一个 SpringApplication 对象实例，然后调用 SpringApplication 对象实例的 run 方法。

（4）查询和加载所有的 SpringApplicationListener 监听器。

（5）SpringApplicationListener 监听器调用其 starting 方法，Spring Boot 通知这些 SpringApplicationListener 监听器，我马上要开始执行了。

（6）创建和准备 Spring Boot 应用将要使用的 Environment 环境，包括配置要使用的 PropertySource 以及 Profile。

（7）创建和初始化应用上下文 ApplicationContext。这一步只是准备工作，并未开始正式创建。

（8）这一步是最重要的，Spring Boot 会通过 @EnableAutoConfiguration 获取所有配置以及其他形式的 IoC 容器配置加载到已经准备完毕的 ApplicationContext。

（9）主要是调用 ApplicationContextInitializer 类的 initialize 方法对应用上下文进行设置和处理。

（10）调用 ApplicationContext 上下文的 refresh 方法，使 Ioc 容器达到可用状态。

（11）查找当前 ApplicationContext 上下文是否注册有 ApplicationRunner 与 CommandLineRunner，如果有，循环遍历执行 ApplicationRunner 和 CommandLineRunner 的 run 方法。

（12）执行 SpringApplicationListener 的 finished 方法，Spring Boot 应用启动完毕。

20.3　spring-boot-starter 原理

20.3.1　自动配置条件依赖

在之前的章节中，我们在项目的 pom 文件中引入了很多 spring-boot-starter 依赖，比如 spring-boot-starter-jdbc、spring-boot-starter-jdbc-logging、spring-boot-starter-web 等。这些带有 spring-boot-starter 前缀的依赖都叫作 Spring Boot 起步依赖，它们有助于 Spring Boot 应用程序的构建。

如果没有起步依赖，假如要使用 Spring MVC 的，我们根本记不住 Spring MVC 到底要引入那些依赖包、到底要使用哪个版本的 Spring MVC、Spring MVC 的 Group 和 Artifact ID 又是多少？

Spring Boot 通过提供众多起步依赖降低项目依赖的复杂度。起步依赖本质上是一个 Maven 项目对象模型，定义了对其他库的传递依赖，这些依赖的合集可以对外提供某项功能。起步依赖的命名表明它们提供某种或某类功能。例如，spring-boot-starter-jdbc 表示提供 JDBC 相关的功能，spring-boot-starter-jpa 表示提供 JPA 相关的功能，等等。表 20-1 中简单地列举了工作中经常使用的起步依赖。

表 20-1　常用的 spring-boot-starter 起步依赖

名　称	描　述
spring-boot-starter-logging	提供 logging 相关的日志功能
spring-boot-starter-thymeleaf	使用 Thymeleaf 视图构建 MVC Web 应用程序的启动器
spring-boot-starter-parent	常被作为父依赖，提供智能资源过滤、智能的插件设置、编译级别和通用的测试框架等

（续表）

名称	描述
spring-boot-starter-web	使用 Spring MVC 构建 Web，包括 RESTful 应用程序。使用 Tomcat 作为默认的嵌入式容器的启动器
spring-boot-starter-test	支持常规的测试依赖，包括 JUnit、Hamcrest、Mockito 以及 spring-test 模块
spring-boot-starter-jdbc	使用 JDBC 与 Tomcat JDBC 连接池的启动器
spring-boot-starter-data-jpa	使用 Spring 数据 JPA 与 Hibernate 的启动器
spring-boot-starter-data-redis	Redis key-value 数据存储与 Spring Data Redis 和 Jedis 客户端启动器
spring-boot-starter-log4j2	提供 log4j2 相关的日志功能
spring-boot-starter-mail	提供邮件相关的功能
spring-boot-starter-activemq	使用 Apache ActiveMQ 的 JMS 启动器
spring-boot-starter-data-mongodb	使用 MongoDB 面向文档的数据库和 Spring Data MongoDB 的启动器
spring-boot-starter-actuator	提供应用监控与健康相关的功能
spring-boot-starter-security	使用 Spring security 的启动器
spring-boot-starter-dubbo	提供 dubbo 框架相关的功能

事实上，起步依赖和项目里的其他依赖没什么区别。引入起步依赖的同时会引入相关的传递依赖。比如 spring-boot-starter-web 起步依赖会引入 spring-webmvc、jackson-databind、spring-boot-starter-tomcat 等传递依赖。如果不想用 spring-boot-starter-web 引入的 spring-webmvc 传递依赖，可以使用<exclusions>标签来排除传递依赖。具体代码如下：

```
<dependency>
    <groupId>org.springframework.boot</groupId>
    <artifactId>spring-boot-starter-web</artifactId>
    <exclusions>
        <!-- 排查 spring-webmvc -->
        <exclusion>
            <groupId>org.springframework</groupId>
            <artifactId>spring-webmvc</artifactId>
        </exclusion>
    </exclusions>
</dependency>
```

假如 spring-boot-starter-web 引入的传递依赖版本过低，可以在 pom 文件中直接引入所需的版本，告诉 Maven 现在需要这个版本的依赖。

传统的 Spring 应用需要在 application.xml 中配置很多 bean，比如 dataSource 的配置，

transactionManager 的配置等。Spring Boot 是如何帮我们完成这些 bean 的配置的呢？下面我们来分析这个过程。我们以第 10 章引入的 mybatis-spring-boot-starter 依赖为例：

```xml
<dependency>
    <groupId>org.mybatis.spring.boot</groupId>
    <artifactId>mybatis-spring-boot-starter</artifactId>
    <version>2.0.1</version>
</dependency>
```

首先，查看 mybatis-spring-boot-starter 包下的内容，具体如图 20-2 所示。

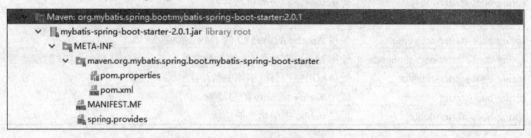

图 20-2　mybatis-spring-boot-starter 包内容

可以看出在 mybatis-spring-boot-starter 包中并没有任何源代码，只有一些配置文件，例如，pom.xml 文件，它的作用就是帮我们引入了相关的 jar 包。在 pom.xml 中可以看出，mybatis-spring-boot-starter 包帮我们引入了 mybatis-spring-boot-autoconfigure 这个包，具体代码如下：

```xml
<?xml version="1.0" encoding="UTF-8"?>
<project xmlns="http://maven.apache.org/POM/4.0.0" xmlns:xsi=
"http://www.w3.org/2001/XMLSchema-instance" xsi:schemaLocation=
"http://maven.apache.org/POM/4.0.0 http://maven.apache.org/xsd/maven-4.0.0.xsd">
  <modelVersion>4.0.0</modelVersion>
  <parent>
    <groupId>org.mybatis.spring.boot</groupId>
    <artifactId>mybatis-spring-boot</artifactId>
    <version>2.0.1</version>
  </parent>
  <artifactId>mybatis-spring-boot-starter</artifactId>
  <name>mybatis-spring-boot-starter</name>
  <properties>
    <module.name>org.mybatis.spring.boot.starter</module.name>
```

```xml
    </properties>
    <dependencies>

        <groupId>org.mybatis.spring.boot</groupId>
        <artifactId>mybatis-spring-boot-autoconfigure</artifactId>
    </dependency>

    //省略代码
    </dependencies>
</project>
```

查看 mybatis-spring-boot-autoconfigure 包下的内容，如图 20-3 所示。

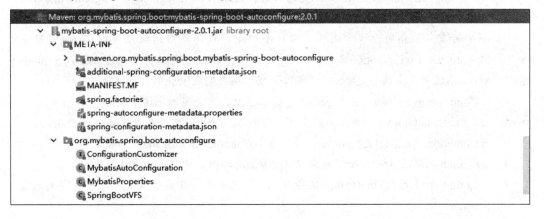

图 20-3　mybatis-spring-boot-autoconfigure 包内容

再查看 MybatisAutoConfiguration 源代码：

```
@Configuration
@ConditionalOnClass({SqlSessionFactory.class, SqlSessionFactoryBean.class})
@ConditionalOnSingleCandidate(DataSource.class)
@EnableConfigurationProperties({MybatisProperties.class})
@AutoConfigureAfter({DataSourceAutoConfiguration.class})
public class MybatisAutoConfiguration implements InitializingBean {
    private static final Logger logger = LoggerFactory.getLogger
(MybatisAutoConfiguration.class);
    private final MybatisProperties properties;
    private final Interceptor[] interceptors;
    private final ResourceLoader resourceLoader;
    private final DatabaseIdProvider databaseIdProvider;
```

```
    private final List<ConfigurationCustomizer> configurationCustomizers;

    @Bean
    @ConditionalOnMissingBean
    public SqlSessionFactory sqlSessionFactory(DataSource dataSource) throws Exception

    //省略代码
}
```

- @Configuration、@Bean：这两个注解一起使用就可以创建一个基于Java代码的配置类。可以把MybatisAutoConfiguration类想象成一份XML配置文件。@Configuration注解的类可以看作Bean实例的工厂，能生产让Spring IoC容器管理的Bean。@Bean注解告诉Spring，一个带有@Bean的注解方法将返回一个对象，该对象应该被注册到Spring容器中。MybatisAutoConfiguration类能自动生成SqlSessionFactory、SqlSessionTemplate等MyBatis的重要实例并交给Spring容器管理，从而完成Bean的自动注册。
- @ConditionalOnClass：某个class位于类路径上，才会实例化这个Bean。
- @ConditionalOnBean：仅在当前上下文中存在某个bean时，才会实例化这个Bean。
- @ConditionalOnSingleCandidate：类似于@ConditionalOnBean。
- @ConditionalOnExpression：当表达式为true的时候，才会实例化这个Bean。
- @ConditionalOnMissingBean：仅在当前上下文中不存在某个Bean时，才会实例化这个Bean。
- @ConditionalOnMissingClass：某个class在类路径上不存在的时候，才会实例化这个Bean。
- @ConditionalOnNotWebApplication：不是Web应用时才会实例化这个Bean。
- @AutoConfigureAfter：在某个Bean完成自动配置后实例化这个Bean。
- @AutoConfigureBefore：在某个Bean完成自动配置前实例化这个Bean。

可见，要完成 MyBatis 的自动配置，需要在类路径中存在 SqlSessionFactory.class、SqlSessionFactoryBean.class 这两个类，需要存在 DataSource 这个 Bean 且这个 Bean 完成自动注册。

进入 DataSourceAutoConfiguration 类，可以看到该类属于 spring-boot-autoconfigure 自动配置包，自动配置这个包帮我们引入了 jdbc、kafka、logging、mail、mongo 等包。很多包需要我们引入相应的 jar 后自动配置才生效。

20.3.2 Bean 参数获取

我们已经知道了 Bean 的配置过程，但是还没有看到 Spring Boot 是如何读取 yml 或者 properites 配置文件的属性来创建数据源，在 DataSourceAutoConfiguration 类里面，使用了 EnableConfigurationProperties 注解，参见如下代码：

```
@Configuration
@ConditionalOnClass({DataSource.class, EmbeddedDatabaseType.class})
@EnableConfigurationProperties({DataSourceProperties.class})
@Import({DataSourcePoolMetadataProvidersConfiguration.class,
DataSourceInitializationConfiguration.class})
public class DataSourceAutoConfiguration {
    public DataSourceAutoConfiguration() {
    }

    //省略代码
}
```

@EnableConfigurationProperties 注解的作用是使@ConfigurationProperties 注解生效。
DataSourceProperties 中封装了数据源的各个属性，且使用了注解@ConfigurationProperties 指定了配置文件的前缀，参见如下代码：

```
@ConfigurationProperties(
    prefix = "spring.datasource"
)
public class DataSourceProperties implements BeanClassLoaderAware,
InitializingBean {
    private ClassLoader classLoader;
    private String name;
    private boolean generateUniqueName;
    private Class<? extends DataSource> type;
    private String driverClassName;
    private String url;
    private String username;
    private String password;
    private String jndiName;
    private DataSourceInitializationMode initializationMode;
```

```
    private String platform;
    private List<String> schema;
    private String schemaUsername;
    private String schemaPassword;
    private List<String> data;
    private String dataUsername;
    private String dataPassword;
    private boolean continueOnError;

//省略代码
}
```

@ConfigurationProperties 注解的作用是把 yml 或者 properties 配置文件转化为 Bean，通过这种方式，把 yml 或者 properties 配置参数转化为 Bean。

20.3.3　Bean 的发现与加载

Spring Boot 默认扫描启动类所在的包下的主类与子类的所有组件，但并没有包括依赖包中的类，那么依赖包中的 Bean 是如何被发现和加载的呢？

我们通常在启动类中加@SpringBootApplication 注解，查看如下源代码：

```
@Target({ElementType.TYPE})
@Retention(RetentionPolicy.RUNTIME)
@Documented
@Inherited
@SpringBootConfiguration
@EnableAutoConfiguration
@ComponentScan(
    excludeFilters = {@Filter(
    type = FilterType.CUSTOM,
    classes = {TypeExcludeFilter.class}
), @Filter(
    type = FilterType.CUSTOM,
    classes = {AutoConfigurationExcludeFilter.class}
)}
)
public @interface SpringBootApplication {
```

```
    //省略代码
}
```

@SpringBootConfiguration 是进入@ SpringBootConfiguration 注解的源代码，你会发现它其实和@Configuration 注解的功能是一样的，只是换了一个名字而已。@ SpringBootConfiguration 的源代码如下：

```
@Target({ElementType.TYPE})
@Retention(RetentionPolicy.RUNTIME)
@Documented
@Configuration
public @interface SpringBootConfiguration {

}
```

- @EnableAutoConfiguration：这个注解的功能非常重要，它用于借助@Import的支持，收集和注册依赖包中相关的Bean定义。
- @ComponentScan：该注解的作用是自动扫描并加载符合条件的组件，比如@Component 和@Repository等，最终将这些Bean定义加载到Spring容器中。

@EnableAutoConfiguration 的源代码如下：

```
@Target({ElementType.TYPE})
@Retention(RetentionPolicy.RUNTIME)
@Documented
@Inherited
@AutoConfigurationPackage
//重要
@Import({AutoConfigurationImportSelector.class})
public @interface EnableAutoConfiguration {

}
```

@EnableAutoConfiguration 注解引入了@AutoConfigurationPackage 和@Import 这两个注解。@AutoConfigurationPackage 的作用是自动配置包，@Import 则是导入需要自动配置的组件。

进入@AutoConfigurationPackage，发现其也是引入了@Import 注解，参见代码如下：

```
@Target({ElementType.TYPE})
@Retention(RetentionPolicy.RUNTIME)
```

```
@Documented
@Inherited
//重要
@Import({Registrar.class})
public @interface AutoConfigurationPackage {
}
```

查看@Import 注解中的 Registrar 类源代码:

```
static class Registrar implements ImportBeanDefinitionRegistrar,
DeterminableImports {
        Registrar() {
        }

        public void registerBeanDefinitions(AnnotationMetadata metadata,
BeanDefinitionRegistry registry) {
            AutoConfigurationPackages.register(registry, (new
AutoConfigurationPackages.PackageImport(metadata)).getPackageName());
        }

        public Set<Object> determineImports(AnnotationMetadata metadata) {
            return Collections.singleton(new
AutoConfigurationPackages.PackageImport(metadata));
        }
    }
```

new AutoConfigurationPackages.PackageImport(metadata)).getPackageName() 和 new AutoConfigurationPackages.PackageImport(metadata)的作用就是加载启动类所在的包下的主类与子类的所有组件注册到 Spring 容器，Spring Boot 默认扫描启动类所在的包下的主类与子类的所有组件。

继续查看 AutoConfigurationImportSelector 类的源代码：

```
public class AutoConfigurationImportSelector implements DeferredImportSelector,
BeanClassLoaderAware, ResourceLoaderAware, BeanFactoryAware, EnvironmentAware,
Ordered {
    private static final AutoConfigurationImportSelector.
AutoConfigurationEntry EMPTY_ENTRY = new AutoConfigurationImportSelector.
AutoConfigurationEntry();
    private static final String[] NO_IMPORTS = new String[0];
```

```java
    private static final Log logger = LogFactory.getLog
(AutoConfigurationImportSelector.class);
    private static final String PROPERTY_NAME_AUTOCONFIGURE_EXCLUDE = 
"spring.autoconfigure.exclude";
    private ConfigurableListableBeanFactory beanFactory;
    private Environment environment;
    private ClassLoader beanClassLoader;
    private ResourceLoader resourceLoader;

    public AutoConfigurationImportSelector() {
    }

    protected List<String> getCandidateConfigurations(AnnotationMetadata 
metadata, AnnotationAttributes attributes) {
        //重要
        List<String> configurations = SpringFactoriesLoader.loadFactoryNames
(this.getSpringFactoriesLoaderFactoryClass(), this.getBeanClassLoader());
        Assert.notEmpty(configurations, "No auto configuration classes found in 
META-INF/spring.factories. If you are using a custom packaging, make sure that file 
is correct.");
        return configurations;
    }

    //省略代码
}
```

SpringFactoriesLoader.loadFactoryNames 方法调用 loadSpringFactories 方法从所有的 jar 包中读取 META-INF/spring.factories 文件信息。loadSpringFactories 的源代码如下：

```java
    public static List<String> loadFactoryNames(Class<?> factoryClass, @Nullable 
ClassLoader classLoader) {
        String factoryClassName = factoryClass.getName();
        return 
(List)loadSpringFactories(classLoader).getOrDefault(factoryClassName, 
Collections.emptyList());
    }

    private static Map<String, List<String>> loadSpringFactories(@Nullable 
ClassLoader classLoader) {
        MultiValueMap<String, String> result = (MultiValueMap)cache.
```

```java
get(classLoader);
            if (result != null) {
                return result;
            } else {
                try {
                    Enumeration<URL> urls = classLoader != null ? classLoader.getResources("META-INF/spring.factories")
                    //加载"META-INF/spring.factories"配置文件中的内容
                     : ClassLoader.getSystemResources("META-INF/spring.factories");
                    LinkedMultiValueMap result = new LinkedMultiValueMap();

                    while(urls.hasMoreElements()) {
                        URL url = (URL)urls.nextElement();
                        UrlResource resource = new UrlResource(url);
                        Properties properties = PropertiesLoaderUtils.loadProperties(resource);
                        Iterator var6 = properties.entrySet().iterator();

                        while(var6.hasNext()) {
                            Entry<?, ?> entry = (Entry)var6.next();
                            String factoryClassName = ((String)entry.getKey()).trim();
                            String[] var9 = StringUtils.commaDelimitedListToStringArray((String)entry.getValue());
                            int var10 = var9.length;

                            for(int var11 = 0; var11 < var10; ++var11) {
                                String factoryName = var9[var11];
                                result.add(factoryClassName, factoryName.trim());
                            }
                        }
                    }

                    cache.put(classLoader, result);
                    return result;
                } catch (IOException var13) {
                    throw new IllegalArgumentException("Unable to load factories from location [META-INF/spring.factories]", var13);
                }
            }
        }
```

下面是 spring-boot-autoconfigure 的 jar 中 spring.factories 文件的部分内容，其中有一个 key 为 org.springframework.boot.autoconfigure.EnableAutoConfiguration 的值定义了需要自动配置的 Bean，通过读取这个配置获取一组@Configuration 类。

```
# Auto Configure
org.springframework.boot.autoconfigure.EnableAutoConfiguration=\
org.springframework.boot.autoconfigure.admin.SpringApplicationAdminJmxAutoConfiguration,\
org.springframework.boot.autoconfigure.aop.AopAutoConfiguration,\
org.springframework.boot.autoconfigure.amqp.RabbitAutoConfiguration,\
org.springframework.boot.autoconfigure.batch.BatchAutoConfiguration,\
org.springframework.boot.autoconfigure.cache.CacheAutoConfiguration,\
org.springframework.boot.autoconfigure.cassandra.CassandraAutoConfiguration,\
org.springframework.boot.autoconfigure.cloud.CloudServiceConnectorsAutoConfiguration,\
org.springframework.boot.autoconfigure.context.ConfigurationPropertiesAutoConfiguration,\
org.springframework.boot.autoconfigure.context.MessageSourceAutoConfiguration,\
org.springframework.boot.autoconfigure.context.PropertyPlaceholderAutoConfiguration,\
org.springframework.boot.autoconfigure.couchbase.CouchbaseAutoConfiguration,\
org.springframework.boot.autoconfigure.dao.PersistenceExceptionTranslationAutoConfiguration,\
org.springframework.boot.autoconfigure.data.cassandra.CassandraDataAutoConfiguration,\
org.springframework.boot.autoconfigure.data.cassandra.CassandraReactiveDataAutoConfiguration,\
```

每个 xxxAutoConfiguration 都是一个基于 Java 的 Bean 配置类。实际上，这些 xxxAutoConfiguration 不是所有都会被加载，会根据 xxxAutoConfiguration 上的@ConditionalOnClass 等条件判断是否加载。通过反射机制将 spring.factories 中的@Configuration 类实例化为对应的 Java 实例。

至此，我们已经知道了怎么发现自动配置的 Bean，最后一步就是怎样将这些 Bean 加载到 Spring 容器中。

将普通类交给 Spring 容器管理，通常有以下方法：

（1）使用 @Configuration 与@Bean 注解。

（2）使用@Controller、@Service、@Repository、@Component 注解标注该类，然后启用@ComponentScan 自动扫描。

（3）使用@Import 方法。

Spring Boot 中采取第 3 种方法，@EnableAutoConfiguration 注解中使用@Import({AutoConfigurationImportSelector.class})注解，AutoConfigurationImportSelector 实现了 DeferredImportSelector 接口，DeferredImportSelector 接口继承了 ImportSelector 接口，ImportSelector 接口只有一个 selectImports 方法。

AutoConfigurationImportSelector 的源代码如下：

```java
public class AutoConfigurationImportSelector implements DeferredImportSelector,
BeanClassLoaderAware, ResourceLoaderAware, BeanFactoryAware, EnvironmentAware,
Ordered {
    private static final
AutoConfigurationImportSelector.AutoConfigurationEntry EMPTY_ENTRY = new
AutoConfigurationImportSelector.AutoConfigurationEntry();
    private static final String[] NO_IMPORTS = new String[0];
    private static final Log logger =
LogFactory.getLog(AutoConfigurationImportSelector.class);
    private static final String PROPERTY_NAME_AUTOCONFIGURE_EXCLUDE =
"spring.autoconfigure.exclude";
    private ConfigurableListableBeanFactory beanFactory;
    private Environment environment;
    private ClassLoader beanClassLoader;
    private ResourceLoader resourceLoader;

    public String[] selectImports(AnnotationMetadata annotationMetadata) {
        if (!this.isEnabled(annotationMetadata)) {
            return NO_IMPORTS;
        } else {
            AutoConfigurationMetadata autoConfigurationMetadata =
AutoConfigurationMetadataLoader.loadMetadata(this.beanClassLoader);
            AutoConfigurationImportSelector.AutoConfigurationEntry
autoConfigurationEntry = this.getAutoConfigurationEntry(autoConfigurationMetadata,
annotationMetadata);
            return StringUtils.toStringArray(autoConfigurationEntry.
getConfigurations());
```

```
        }
    }
}
```
DeferredImportSelector 与 ImportSelector 的源代码如下：
```
//DeferredImportSelector 源代码
public interface DeferredImportSelector extends ImportSelector {
    @Nullable
    default Class<? extends DeferredImportSelector.Group> getImportGroup() {
        return null;
    }
    //省略代码
}

//ImportSelector 源代码
public interface ImportSelector {
    String[] selectImports(AnnotationMetadata var1);
}
```

20.3.4 自定义 starter

Spring Boot 提供的 starter 都是以 spring-boot-starter-xxx 的方式命名的，针对自定义的 starter，官方建议以 xxx-spring-boot-starter 命名予以区分。

首先，创建一个 Maven 项目，具体步骤如下：

步骤 01 单击菜单栏的【File】→【New】→【Project】，如图 20-4 所示。

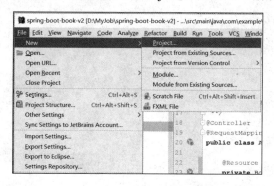

图 20-4 创建 Maven 项目

步骤 02 在打开的【New Project】对话框中选择 Maven，勾选 Create from archetype，选择 maven-archetype-quickstart 选项，如图 20-5 所示。

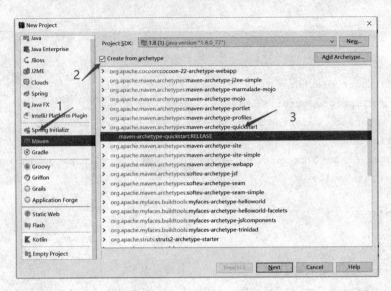

图 20-5　New Project 弹出框

步骤 03　输入 GroupId 和 ArtifactId，例如，GroupId 输入框输入"org.springframework.boot"，在 ArtifactId 输入框中输入"ay-spring-boot-starter"，如图 20-6 所示。

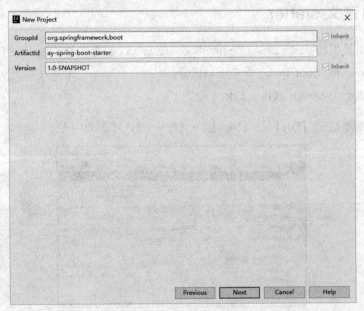

图 20-6　输入 GroupId 和 ArtifactId

步骤 04　选择 Maven 的配置，这里笔者选择自己下载的 Maven，也可以使用默认选项，如图 20-7 所示。

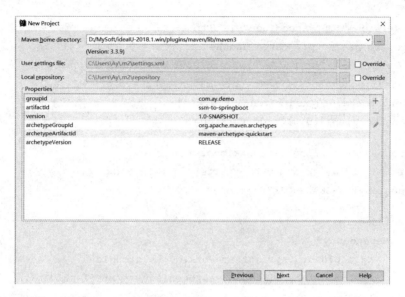

图 20-7　选择 Maven home 地址

步骤 05　输入项目名称和项目存放目录，最后单击【Finish】按钮，如图 20-8 所示。

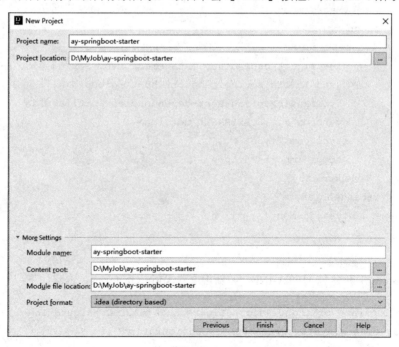

图 20-8　输入项目名称和地址

Maven 项目创建完成后，在 pom.xml 文件中添加依赖，具体代码如下：

```xml
<?xml version="1.0" encoding="UTF-8"?>
<project xmlns="http://maven.apache.org/POM/4.0.0"
         xmlns:xsi="http://www.w3.org/2001/XMLSchema-instance"
         xsi:schemaLocation="http://maven.apache.org/POM/4.0.0 http://maven.apache.org/xsd/maven-4.0.0.xsd">
    <modelVersion>4.0.0</modelVersion>

    <groupId>org.springframework.boot</groupId>
    <artifactId>ay-spring-boot-starter</artifactId>
    <version>1.0-SNAPSHOT</version>

    <dependencies>
        <dependency>
            <groupId>org.springframework.boot</groupId>
            <artifactId>spring-boot-autoconfigure</artifactId>
        </dependency>
    </dependencies>

    <dependencyManagement>
        <dependencies>
            <dependency>
                <!-- Import dependency management from Spring Boot -->
                <groupId>org.springframework.boot</groupId>
                <artifactId>spring-boot-dependencies</artifactId>
                <version>2.1.6.RELEASE</version>
                <type>pom</type>
                <scope>import</scope>
            </dependency>
        </dependencies>
    </dependencyManagement>

</project>
```

在 resources 下新建包 META-INF，并新增文件 spring.factories，内容如下：

```
org.springframework.boot.autoconfigure.EnableAutoConfiguration=\
  com.ay.config.AyStarterEnableAutoConfiguration
```

其中，AyStarterEnableAutoConfiguration 为自动配置的核心类，Spring Boot 会扫描该文件作为配置类。

在 src\main\java 目录下创建 com.ay.config 包，并在该包下创建 AyStarterEnableAutoConfiguration.java 类，具体代码如下：

```java
/**
 * @author Ay
 * @create 2019/09/08
 **/

@Configuration
@ConditionalOnClass(HelloService.class)
@EnableConfigurationProperties(HelloServiceProperties.class)
public class AyStarterEnableAutoConfiguration {

    private final HelloServiceProperties helloServiceProperties;

    @Autowired
    public AyStarterEnableAutoConfiguration(HelloServiceProperties helloServiceProperties) {
        this.helloServiceProperties = helloServiceProperties;
    }

    @Bean
    @ConditionalOnProperty(prefix = "hello.service", name = "enable", havingValue = "true")
    HelloService helloService() {
        return new HelloService(helloServiceProperties.getPrefix(), helloServiceProperties.getSuffix());
    }
}
```

@Configuration、@ConditionalOnClass(HelloService.class) 和 @EnableConfigurationProperties 注解的含义在之前章节都提到过，这里不再赘述。

HelloServiceProperties 的代码如下：

```java
/**
 * @author Ay
 * @create 2019/09/08
 **/
@ConfigurationProperties("hello.service")
```

```java
public class HelloServiceProperties {

    private String prefix;

    private String suffix;
    //省略 set、get 方法
}
```

HelloService 的代码如下：

```java
/**
 * @author Ay
 * @create 2019/09/08
 **/
public class HelloService {
    private String prefix;
    private String suffix;
    public HelloService(String prefix, String suffix) {
        this.prefix = prefix;
        this.suffix = suffix;
    }

    public String say(String text) {
        return String.format("%s , hi , %s , %s", prefix, text, suffix);
    }
}
```

当项目依赖该 starter 时，并且配置文件中包含 hello.service 为前缀且 hello.service.enable 为 true 时，就会自动生成 HelloService 的 Bean。

最后，我们可以创建一个新的 Spring Boot 项目，并在项目的 pom.xml 文件中引入自定义 starter 依赖：

```xml
<dependency>
        <groupId>org.springframework.boot</groupId>
        <artifactId>ay-spring-boot-starter</artifactId>
        <version>1.0-SNAPSHOT</version>
</dependency>
```

在 application.properties 配置文件中添加如下配置：

```
hello.service.prefix=pre
hello.service.suffix=suf
hello.service.enable=true
```

这样,就可以在新的 Spring Boot 项目中使用 HelloService 类并调用 say 方法了。

20.4　跨域访问

对于前后端分离项目,如果前端项目与后端项目部署在两个不同的域下,那么势必会引起跨域问题的出现。针对跨域问题,第一个想到的解决方案可能就是使用 jsonp,但是 jsonp 方式有一些不足,即 jsonp 方式只能通过 get 请求方式来传递参数,当然还有其他的不足之处。在 Spring-Boot 中可以通过 CORS(Cross-Origin Resource Sharing,跨域资源共享)协议来解决跨域问题,CORS 是一个 W3C 标准,它允许浏览器向不同源的服务器发出 xmlHttpRequest 请求,我们可以继续使用 Ajax 进行请求访问。Spring-MVC 4.2 版本增加了对 CORS 的支持,具体做法如下:

```
@Configuration
public class MyWebAppConfigurer extends WebMvcConfigurerAdapter{

    @Override
    public void addCorsMappings(CorsRegistry registry) {
        registry.addMapping("/**");
    }
}
```

可以在 addMapping 方法中配置路径,/** 代表所有路径。当然,也可以修改其他的属性,例如:

```
@Configuration
public class MyWebAppConfigurer extends WebMvcConfigurerAdapter{

    @Override
    public void addCorsMappings(CorsRegistry registry) {
        registry.addMapping("/api/**")
            .allowedOrigins("http://192.168.1.97")
            .allowedMethods("GET", "POST")
```

```
            .allowCredentials(false).maxAge(3600);
    }
}
```

以上两种方式都是针对全局配置。如果想做到更细致，可以在 Controller 类中使用 @CrossOrigin 注解，代码如下：

```
@CrossOrigin(origins = "http://192.168.1.97:8080", maxAge = 3600)
@RequestMapping("rest_index")
@RestController
public class AyController{}
```

这样，就可以指定 AyController 中的所有方法都能处理来自 http:19.168.1.97:8080 的请求。

20.5 优雅关闭

什么叫优雅关闭？简单地说，就是对应用进程发送停止指令之后，能保证正在执行的业务操作不受影响。应用接收到停止指令之后，停止接收新请求，但可以继续完成已有请求的处理。

20.5.1 Java 优雅停机

Java 底层是支持优雅停机的，Java 底层能够捕获到操作系统的 SIGINT/SIGTERM 停止指令，通过 Runtime.getRuntime().addShutdownHook() 向 JVM 注册 Shutdown hook 线程，当 JVM 收到停止信号后，该线程将被激活运行，这时就可以向其他线程发出中断指令，进而快速而优雅地关闭整个程序。具体实例如下：

```
/**
 * 描述：Java 优雅关闭
 * @author Ay
 * @create 2019/09/01
 **/
public class JavaShutdownTest {

    public static void main(String[] args) {
        System.out.println("step-1: main thread start");
        Thread mainThread = Thread.currentThread();
```

```java
        //注册一个 ShutdownHook
        ShutdownSampleHook thread=new ShutdownSampleHook(mainThread);
        Runtime.getRuntime().addShutdownHook(thread);
        try {
            //主线程sleep 30s
            Thread.sleep(30*1000);
        } catch (InterruptedException e) {
            System.out.println("step-3: mainThread get interrupt signal.");
        }
        System.out.println("step-4: main thread end");
    }
}

/**
 * 钩子
 */
class ShutdownSampleHook extends Thread {

    //主线程
    private Thread mainThread;

    @Override
    public void run() {
        System.out.println("step-2: shut down signal received.");
//给主线程发送一个中断信号
        mainThread.interrupt();
        try {
            mainThread.join(); //等待 mainThread 正常运行完毕
        } catch (InterruptedException e) {
            e.printStackTrace();
        }
        System.out.println("step-5: shut down complete.");
    }

    /**
     * 构造方法
     * @param mainThread
     */
    public ShutdownSampleHook(Thread mainThread) {
```

```
        this.mainThread=mainThread;
    }
}
```

- mainThread.interrupt()：该方法将给主线程发送中断信号。如果主线程没有进入阻塞状态，interrupt()不能起什么作用；如果主线程处于阻塞状态，该线程将得到一个InterruptedException 异常。
- mainThread.join()：等待 mainThread 正常运行完毕。

在命令行窗口执行 JavaShutdownTest 类，具体代码如下：

在项目的目录 D:\MyJob\spring-boot-book-v2\target\classes 下执行 java 命令

同时指定 classpath 目录，程序运行结果如下：

```
D:\MyJob\spring-boot-book-v2\target\classes>java -classpath
D:\MyJob\spring-boot-book-v2\target\classes
com.example.demo.shutdown.JavaShutdownTest
    step-1: main thread start
    step-4: main thread end
    step-2: shut down signal received.
    step-5: shut down complete.
```

再次运行命令，此时在程序运行过程中，按下 **Ctrl+C** 键程序很快就结束了，最终的输出是：

程序执行过程中，按下 Ctrl+C 键，以下是程序打印的信息

```
D:\MyJob\spring-boot-book-v2\target\classes>java -classpath
D:\MyJob\spring-boot-book-v2\target\classes
com.example.demo.shutdown.JavaShutdownTest
    step-1: main thread start
    step-2: shut down signal received.
    step-3: mainThread get interrupt signal.
    step-4: main thread end
    step-5: shut down complete.
```

从运行实例得知，我们简单地实现了 Java 应用程序的优雅退出。

这里补充一下 Linux kill 命令的知识点。在 Linux 操作系统中，kill 命令用于终止指定的进程。kill 命令常用的信号选项如下：

（1）kill -2   pid：向指定 pid（进程号）发送 SIGINT 中断信号，等同于 Ctrl+C 键。
（2）kill -9   pid：向指定 pid 发送 SIGKILL 立即终止信号。
（3）kill -15   pid：向指定 pid 发送 SIGTERM 终止信号。
（4）kill   pid：等同于 kill 15 pid。

下面是常用的信号：

```
SIGHUP      1  终端断线
SIGINT      2  中断（同 Ctrl + C），信号会被当前进程树接收到,而且它的子进程也会收到
SIGQUIT     3  退出（同 Ctrl + \）
SIGKILL     9  强制终止，程序不能捕获该信号，最粗暴最快速结束程序的方法
SIGTERM     15 终止（正常结束），缺省信号，信号会被当前进程接收到，但它的子进程不会收到，如
果当前进程被 kill 掉，它的子进程的父进程将变成 init 进程（init 进程是 pid 为 1 的进程）
SIGCONT     18 继续（同 fg/bg 命令）
SIGSTOP     19 停止
SIGTSTP     20 暂停（同 Ctrl + Z）
```

20.5.2　Spring Boot 优雅停机

Java Web 服务器通常也支持优雅退出，例如 tomcat 提供了如下命令：

```
### 先等 n 秒后，然后停止 tomcat
catalina.sh stop n
### 先等 n 秒后，然后 kill -9 tomcat
catalina.sh stop n -force
```

如果 Spring Boot Web 项目使用外置 tomcat，可直接使用上面的 tomcat 命令完成优雅停机。但通常使用的是内置 tomcat 服务器，这时就需要编写代码来支持优雅停止。

可能有些读者会问，Spring Boot 的 Actuator 端点不是提供了 shutdown 优雅停机功能吗？官方文档也是这么宣传的，其实并没有实现优雅停机功能。

下面简单实现一下 Spring Boot 的优雅关闭代码，我们增加 GracefulShutdown 监听类，当 tomcat 收到 kill 信号后，应用程序先关闭新的请求，然后等待 30 秒，最后结束整个程序。

```
import org.apache.catalina.connector.Connector;
import org.slf4j.Logger;
import org.slf4j.LoggerFactory;
import org.springframework.boot.web.embedded.tomcat.TomcatConnectorCustomizer;
```

```java
import org.springframework.context.ApplicationListener;
import org.springframework.context.event.ContextClosedEvent;
import java.util.concurrent.Executor;
import java.util.concurrent.ThreadPoolExecutor;
import java.util.concurrent.TimeUnit;

/**
 * 描述：优雅关闭
 * @author ay
 * @date 2019-09-01
 */
public class GracefulShutdown implements TomcatConnectorCustomizer, ApplicationListener<ContextClosedEvent> {
    private static final Logger log = LoggerFactory.getLogger(GracefulShutdown.class);
    private volatile Connector connector;

    @Override
    public void customize(Connector connector) {
        this.connector = connector;
    }

    /**
     * 描述：监听上下文关闭事件
     * @param event
     */
    @Override
    public void onApplicationEvent(ContextClosedEvent event) {
        Executor executor = this.connector.getProtocolHandler().getExecutor();
        if (executor instanceof ThreadPoolExecutor) {
            try {
                //不再接受新的线程，并且等待之前提交的线程都执行完后再关闭
                ThreadPoolExecutor threadPoolExecutor = (ThreadPoolExecutor) executor;
                threadPoolExecutor.shutdown();
                if (!threadPoolExecutor.awaitTermination(30, TimeUnit.SECONDS)) {
                    log.warn("Tomcat thread pool did not shut down gracefully within "
                            + "30 seconds. Proceeding with forceful shutdown");
                }
```

```
            } catch (InterruptedException ex) {
                Thread.currentThread().interrupt();
            }
        }
    }
}
```

- Connector：用于接收请求并将请求封装成Request和Response来进行具体的处理，最底层使用socket来进行连接。Request和Response是按照HTTP协议封装的，所以Connector同时实现了TCP/IP协议和HTTP协议，Request和Response封装完成之后交给Container来进行处理，Container就是Servlet的容器，Container处理完成后返回给Connector，最后Connector使用Socket将结果返回客户端，请求完成。
- ProtocolHandler：Connector中具体使用ProtocolHandler来处理请求，不同的ProtocolHandler代表不同的连接类型。例如，Http1Protocol使用普通的Sockct连接，而Http1BioProtocol则使用NioSocket来连接。
- Executor：Connector建立连接后，服务器会分配一个线程（可能是从线程池）去服务这个连接，即执行doPost等方法。执行完回收线程。显然这一步是一个同步的过程，tomcat对应的是Executor。
- ThreadPoolExecutor.shutdown()：不再接受新的线程，并且等待之前提交的线程都执行完后，再关闭。

在@SpringBootApplication入口类中，增加下面的代码，注册之前定义的Connector监听器：

```
@Bean
public GracefulShutdown gracefulShutdown() {
    return new GracefulShutdown();
}

@Bean
public ConfigurableServletWebServerFactory webServerFactory(final GracefulShutdown gracefulShutdown) {
    TomcatServletWebServerFactory factory = new TomcatServletWebServerFactory();
    factory.addConnectorCustomizers(gracefulShutdown);
    return factory;
}
```

该方案的代码来自官方issue中的讨论，添加这些代码到Spring Boot项目中，再重新启动之后发起测试请求，然后发送 kill 停止指令（kill -2【Ctrl + C】、kill -15）。测试结果是正在执行的操作不会终止，直到执行完成，不再接收新的请求，最后正常终止进程（业务执行完成后，进程立即停止）。

20.6 将 SSM/Maven 项目改造为 Spring Boot 项目

Spring Boot 项目未出现之前，大部分项目都采用 SSM（Spring + Spring MVC + MyBatis）架构，如何把 SSM 项目或者 Maven 类型的项目改造成 Spring Boot 项目，是我们这一节要讨论的重点。

20.6.1 创建 Maven 项目

首先，创建一个 Maven 项目，具体步骤如下。

步骤 01 单击菜单栏的【File】→【New】→【Project】，在【New Project】对话框中选择 Maven，勾选 Create from archetype 复选框，选择 maven-archetype-quickstart 选项，如图 20-9 所示。

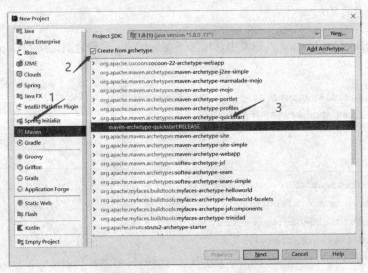

图 20-9 New Project 对话框

步骤 02 输入 GroupId 和 ArtifactId，例如，在 GroupId 输入框中输入"com.ay.demo"，在 ArtifactId 输入框中输入"ssm-to-springboot"，如图 20-10 所示。

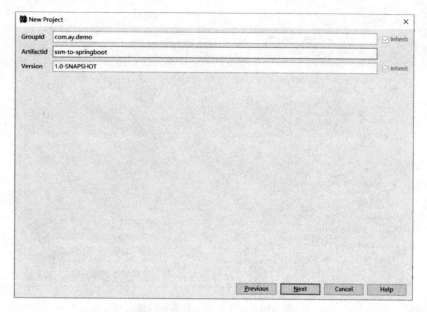

图 20-10　输入 GroupId 和 ArtifactId

步骤 03 选择 Maven 的配置，这里笔者选择自己下载的 Maven，也可以选择默认选项，如图 20-11 所示。

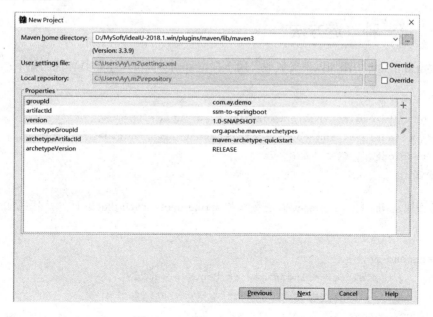

图 20-11　选择 Maven home 地址

步骤 04 输入项目名称和项目存放目录，最后单击【Finish】按钮，如图 20-12 所示。

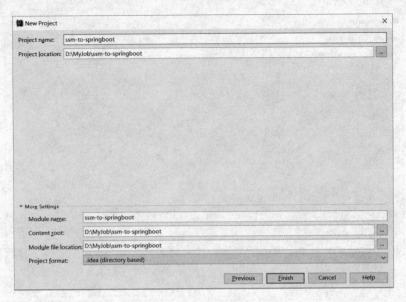

图 20-12　输入项目名称和地址

Maven 项目创建完成后，把它当作一个 SSM 架构的项目，现在我们把 Maven 项目改造成一个 Spring Boot 项目。

20.6.2　第一种改造方法

在 pom.xml 文件中添加以下父依赖，用于指定 spring-boot 版本，具体代码如下：

```
<parent>
    <groupId>org.springframework.boot</groupId>
    <artifactId>spring-boot-starter-parent</artifactId>
    <version>2.1.6.RELEASE</version>
</parent>
```

同时添加 spring-boot-starter-web 依赖包和 spring-boot-maven-plugin 插件，代码如下：

```
<dependencies>
    <dependency>
        <groupId>org.springframework.boot</groupId>
        <artifactId>spring-boot-starter-web</artifactId>
    </dependency>
</dependencies>
<build>
```

```xml
        <plugins>
            <plugin>
                <groupId>org.springframework.boot</groupId>
                <artifactId>spring-boot-maven-plugin</artifactId>
            </plugin>
        </plugins>
</build>
```

pom.xml 文件修改完成后，开发启动类，具体代码如下：

```java
/**
 * 描述：入口类
 * @author Ay
 * @create 2019/09/01
 **/
@SpringBootApplication
public class ApplicationStartup {

    public static void main(String[] args) {
        SpringApplication.run(ApplicationStartup.class, args);
    }
}
```

最后，在 src\main\resources 目录下创建 application.properties 配置文件，配置文件不需要添加任何内容。自此，SSM 项目改造成 Spring Boot 项目已完成，运行启动类的 main 方法，便可以启动项目。当然，该 SSM 项目非常简单，真实的 SSM 项目会有更复杂的依赖关系。

20.6.3 第二种改造方法

通常，在我们自己的项目中往往会定义公司自己的 parent 依赖，这种情况下，上面的做法就行不通了。那么，该如何来解决这个问题呢？其实，在 Spring Boot 的官方网站中也给出了变通的方法，下面我们来看看具体的做法。

修改 pom.xml 配置文件，完整的 pom.xml 文件代码如下：

```xml
<?xml version="1.0" encoding="UTF-8"?>
<project xmlns="http://maven.apache.org/POM/4.0.0"
         xmlns:xsi="http://www.w3.org/2001/XMLSchema-instance"
         xsi:schemaLocation="http://maven.apache.org/POM/4.0.0
```

```xml
http://maven.apache.org/xsd/maven-4.0.0.xsd">
    <modelVersion>4.0.0</modelVersion>

    <groupId>com.ay.demo</groupId>
    <artifactId>ssm-to-springboot</artifactId>
    <version>1.0-SNAPSHOT</version>
    <dependencies>
        <!-- spring boot 依赖 -->
        <dependency>
            <!-- Import dependency management from Spring Boot -->
            <groupId>org.springframework.boot</groupId>
            <artifactId>spring-boot-dependencies</artifactId>
            <version>2.1.6.RELEASE</version>
            <type>pom</type>
            <scope>import</scope>
        </dependency>
        <!-- spring boot web 依赖 -->
        <dependency>
            <groupId>org.springframework.boot</groupId>
            <artifactId>spring-boot-starter-web</artifactId>
            <version>2.1.6.RELEASE</version>
        </dependency>
    </dependencies>
    <!-- 用了重新打包应用程序 -->
    <build>
        <plugins>
            <plugin>
                <groupId>org.springframework.boot</groupId>
                <artifactId>spring-boot-maven-plugin</artifactId>
                <version>1.5.7.RELEASE</version>
                <configuration>
                    <executable>true</executable>
                </configuration>
                <executions>
                    <execution>
                        <goals>
                            <goal>repackage</goal>
                        </goals>
```

```
                </execution>
            </executions>
        </plugin>
    </plugins>
</build>
</project>
```

此时，重新运行启动类的 main 方法，便可以启动项目。同时，在 target 目录中可以看到 ssm-to-springboot 项目被打包成了一个可执行的 jar 文件，如图 20-13 所示。

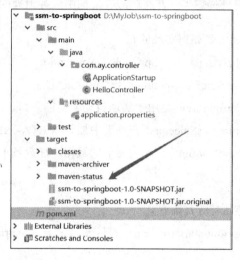

图 20-13　ssm-to-springboot 打包成可运行 jar

更多内容请参考官方网址：

https://docs.spring.io/spring-boot/docs/current/reference/htmlsingle/#using-boot-maven-without-a-parent

20.7　思考题

1. Spring、Spring Boot 和 Spring Cloud 的关系是什么？

答：Spring 最初的两大核心功能是 Spring IoC 和 Spring AOP，Spring 在这两大核心功能上不断发展，才有了 Spring 事务、Spring MVC 等一系列伟大的产品，最终成就了 Spring 帝国，到了后期 Spring 几乎可以解决企业开发中的所有问题。

Spring Boot 是在强大的 Spring 帝国生态基础上面发展而来的，发明 Spring Boot 不是为了取代 Spring，是为了让人们更容易地使用 Spring。

Spring Cloud 是一系列框架的有序集合。它利用 Spring Boot 的开发便利性巧妙地简化了分布式系统基础设施的开发，如服务发现注册、配置中心、消息总线、负载均衡、断路器、数据监控等，都可以用 Spring Boot 的开发风格做到一键启动和部署。

Spring Cloud 是为了解决微服务架构中服务治理而提供的一系列功能的开发框架，并且 Spring Cloud 是完全基于 Spring Boot 而开发，Spring Cloud 利用 Spring Boot 的特性整合了开源行业中优秀的组件，整体对外提供了一套在微服务架构中服务治理的解决方案。

2. Spring Boot 的自动配置是如何实现的？

答：Spring Boot 项目的启动注解是@SpringBootApplication，其实它就是由@Configuration、@ComponentScan 和@EnableAutoConfiguration 3 个注解组成的。

其中，@EnableAutoConfiguration 是用于实现自动配置的入口，该注解又通过 @Import 注解导入了 AutoConfigurationImportSelector，在该类中加载 META-INF/spring.factories 的配置信息，然后筛选出以 EnableAutoConfiguration 为 key 的数据，加载到 IoC 容器中，实现自动配置功能。

3. 开启 Spring Boot 特性有哪几种方式？

答：（1）继承 spring-boot-starter-parent 项目；（2）导入 spring-boot-dependencies 项目依赖。

4. 如何理解 Spring Boot 中的 starter？

答：starters 可以理解为启动器，它包含了一系列可以集成到应用里面的依赖包，你可以一站式集成 Spring 及其他技术，而不需要到处找示例代码和依赖包。如你想使用 Spring JPA 访问数据库，只要加入 spring-boot-starter-data-jpa 启动器依赖就能使用了。

5. 如何在 Spring Boot 启动的时候运行一些特定的代码？

答：可以实现接口 ApplicationRunner 或者 CommandLineRunner，这两个接口实现方式一样，它们都只提供了一个 run 方法。

第 21 章

实战高并发秒杀系统

本章主要介绍如何使用 Spring Boot 搭建一个高可用、高性能、高并发的秒杀系统。

21.1 秒杀系统业务

21.1.1 什么是秒杀

秒杀系统是网络商家为了促销等目的进行的网上限时抢购活动。比如淘宝的秒杀、一元抢购以及 12306 的购票等，都属于秒杀系统。用户在规定的时间内，定时定量的秒杀，无论商品是否秒杀完毕，该场次的秒杀活动都会结束。

秒杀系统具有瞬时流量、高并发读、高并发写以及高可用等特点。秒杀时会有大量用户在同一时间进行抢购，瞬时并发访问量突然增加 10 倍，甚至 100 倍以上都有可能。

秒杀系统的架构设计思想主要有：

（1）缓存

把部分业务逻辑迁移到内存的缓存或者 Redis 中，从而极大地提高并发读效率。

（2）削峰

秒杀开始的一瞬间，会有大量用户冲进来，所以在开始时会有一个瞬间流量峰值。如何使瞬间的流量峰值变得更平缓，是成功设计秒杀系统的关键。要实现流量的削峰填谷，一般的方法是采用缓存和 MQ 中间件。

（3）异步

将同步业务设计成异步处理的任务，以提高网站的整体可用性。

（4）限流

由于活动库存量一般都很少，只有少部分用户才能秒杀成功，所以需要限制大部分用户流量，只准少量用户流量进入后端服务器。

21.1.2　秒杀系统的工作流程

秒杀系统的整体工作流程，如图 21-1 所示。

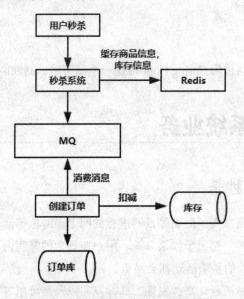

图 21-1　秒杀系统的工作流程图

21.2　秒杀系统的简单实现

21.2.1　创建 Spring Boot 项目

参考 1.3 节内容创建 Spring Boot 项目，项目名称为 speed-kill-system。

21.2.2 库表设计与 Model 实体类

创建数据库 speed-kill-system 并设计秒杀系统的库表，这里的库表分别是 ay_product（商品表）、ay_user（用户表）、ay_user_kill_product（秒杀成功明细表），同时，初始化用户和商品数据。具体的 SQL 语句如下：

```sql
---创建数据库
CREATE DATABASE speed-kill-system;

-- ----------------------------
-- Table structure for ay_product
-- 商品表
-- ----------------------------
DROP TABLE IF EXISTS `ay_product`;
CREATE TABLE `ay_product` (
  `id` bigint(20) NOT NULL COMMENT '商品id',
  `name` varchar(255) NOT NULL DEFAULT '' COMMENT '商品名称',
  `number` int(11) NOT NULL DEFAULT '0' COMMENT '商品数量',
  `start_time` timestamp NOT NULL DEFAULT '0000-00-00 00:00:00'
ON UPDATE CURRENT_TIMESTAMP COMMENT '秒杀开始时间',
  `end_time` timestamp NOT NULL DEFAULT '0000-00-00 00:00:00'
ON UPDATE CURRENT_TIMESTAMP COMMENT '秒杀结束时间',
  `create_time` timestamp NOT NULL DEFAULT '0000-00-00 00:00:00'
ON UPDATE CURRENT_TIMESTAMP COMMENT '创建时间',
  `product_img` varchar(255) DEFAULT NULL COMMENT '商品图片',
  PRIMARY KEY (`id`),
  KEY `idx_name` (`name`) USING BTREE
) ENGINE=InnoDB DEFAULT CHARSET=utf8mb4;

-- ----------------------------
-- Records of ay_product
-- 插入 3 条商品记录
-- ----------------------------
INSERT INTO `ay_product` VALUES ('1', '一步一步学 Spring Boot2：微服务项目实战', '100', '2019-08-19 23:05:47', '2019-08-19 23:05:47', '2019-08-19 23:05:47', '/spring-boot.jpg');
INSERT INTO `ay_product` VALUES ('2', 'Spring MVC+Mybatis：快速开发与实战', '100', '2019-08-19 23:05:55', '2019-08-19 23:05:55', '2019-08-19 23:05:55',
```

```sql
'/spring-mvc.jpg');
INSERT INTO `ay_product` VALUES ('3', 'Spring 5企业级开发实战', '100',
'2019-08-19 23:06:15', '2019-08-19 23:06:15', '2019-08-19 23:06:15',
'/spring-5.jpg');

-- ----------------------------
-- Table structure for ay_user
-- 用户表
-- ----------------------------
DROP TABLE IF EXISTS `ay_user`;
CREATE TABLE `ay_user` (
  `id` bigint(20) NOT NULL COMMENT '用户id',
  `name` varchar(255) NOT NULL COMMENT '用户名',
  `phone_number` varchar(11) DEFAULT NULL COMMENT '电话号码',
  PRIMARY KEY (`id`)
) ENGINE=InnoDB DEFAULT CHARSET=utf8mb4;

-- ----------------------------
-- Records of ay_user
-- 插入2条用户记录
-- ----------------------------
INSERT INTO `ay_user` VALUES ('1', 'ay', '15988888888');
INSERT INTO `ay_user` VALUES ('2', 'al', '15900000000');

-- ----------------------------
-- Table structure for ay_user_kill_product
-- 商品描述记录表
-- ----------------------------
DROP TABLE IF EXISTS `ay_user_kill_product`;
CREATE TABLE `ay_user_kill_product` (
  `product_id` bigint(20) DEFAULT NULL COMMENT '商品id',
  `user_id` bigint(20) DEFAULT NULL COMMENT '用户id',
  `state` tinyint(4) DEFAULT NULL COMMENT '状态, -1: 无效; 0: 成功; 1: 已付款',
  `create_time` timestamp NULL DEFAULT NULL
ON UPDATE CURRENT_TIMESTAMP COMMENT '创建时间',
  `id` bigint(20) NOT NULL AUTO_INCREMENT COMMENT '唯一主键',
  PRIMARY KEY (`id`)
) ENGINE=InnoDB AUTO_INCREMENT=13 DEFAULT CHARSET=utf8mb4;
```

库表创建完成后,接下来创建库表对应的 Model 实体类。

在 src\main\java 目录下创建 com.example.demo.model 包,在包下创建 AyUser.java 实体类,具体代码如下:

```java
/**
 * 描述:用户类
 * @author Ay
 * @create 2019/07/02
 **/
@Entity
@Table(name = "ay_user")
public class AyUser implements Serializable {

    //主键
    @Id
    private Integer id;
    //用户名
    private String name;
    //电话号码
    private String phoneNumber;

    //省略 set、get 方法
}
```

在 com.example.demo.model 包下创建 AyProduct.java 商品类,具体代码如下:

```java
/**
 * 描述:商品类
 * @author Ay
 * @create 2019/08/17
 **/
@Entity
@Table(name = "ay_product")
public class AyProduct implements Serializable {
    /**
     * 商品 id
     */
    @Id
    private Integer id;
```

```java
/**
 * 商品图片
 */
private String productImg;

/**
 * 商品名称
 */
private String name;
/**
 * 商品数量
 */
private Integer number;
/**
 * 秒杀开始时间
 */
private Date startTime;
/**
 * 秒杀结束时间
 */
private Date endTime;
/**
 * 创建时间
 */
private Date createTime;

//省略 set、get 方法
}
```

在 com.example.demo.model 包下创建 AyUserKillProduct.java 商品类，具体代码如下：

```java
/**
 * 描述：用户秒杀商品记录
 * @author Ay
 * @create 2019/08/17
 **/
@Entity
@Table(name = "ay_user_kill_product")
public class AyUserKillProduct implements Serializable {
```

```java
    @Id
    @GeneratedValue(strategy = GenerationType.IDENTITY)
    private Integer id;
    /**
     * 商品 id
     */
    private Integer productId;
    /**
     * 用户 id
     */
    private Integer userId;
    /**
     * 状态, -1: 无效; 0: 成功; 1: 已付款'
     */
    private Integer state;
    /**
     * 创建时间
     */
    private Date createTime;

    //省略 set、get 方法
}
```

- @Id：声明此属性为主键。
- @GeneratedValue：指定主键的生成策略。在 javax.persistence.GenerationType 中定义以下几种可供选择的策略：
    - IDENTITY：采用数据库 ID 自增长的方式来自增主键字段，注意，Oracle 不支持这种方式。
    - AUTO：JPA 自动选择合适的策略，为默认选项。
    - SEQUENCE：通过序列产生主键，通过 @SequenceGenerator 注解指定序列名，注意，MySQL 不支持这种方式。
    - TABLE：通过表产生主键，框架通过表模拟序列产生主键，使用该策略可以使应用更易于数据库移植。

关于 SEQUENCE 和 TABLE 等主键生成策略，读者可自己查阅相关资料学习。

最后，在 com.example.demo.model 包下创建枚举类 KillStatus.java，具体代码如下：

```java
/**
 * 描述：秒杀状态类
 * @author Ay
 * @create 2019/08/21
 **/
public enum KillStatus {

    IN_VALID(-1, "无效"),
    SUCCESS(0, "成功"),
    PAY(1,"已付款");

    private int code;
    private String name;

    KillStatus(){}

    KillStatus(int code,String name){
        this.code = code;
        this.name = name;
    }

    public int getCode() {
        return code;
    }

    public void setCode(int code) {
        this.code = code;
    }
}
```

21.2.3 集成 MySQL 和 JPA

在 Spring Boot 中集成 MySQL 和 JPA，首先需要在 pom.xml 文件中引入所需的依赖，具体代码如下：

```xml
<dependency>
    <groupId>org.springframework.boot</groupId>
    <artifactId>spring-boot-starter-data-jpa</artifactId>
</dependency>
```

```xml
<!-- mysql start -->
<dependency>
    <groupId>mysql</groupId>
    <artifactId>mysql-connector-java</artifactId>
</dependency>
<dependency>
    <groupId>org.springframework.boot</groupId>
    <artifactId>spring-boot-starter-jdbc</artifactId>
</dependency>
```

在 application.properties 配置文件中添加 MySQL 配置：

```
### MySQL 连接信息
spring.datasource.url=jdbc:mysql://127.0.0.1:3306/speed-kill-system?serverTimezone=UTC
###用户名
spring.datasource.username=root
###密码
spring.datasource.password=123456
###驱动
spring.datasource.driver-class-name=com.mysql.jdbc.Driver
```

speed-kill-system 数据库上一节已经创建完成，用户名 root，密码 123456。

在 com.example.demo.model 包下创建 ProductRepository.java 类，负责商品的 CRUD（增删改查），具体代码如下：

```java
/**
 * 描述：商品 Repository
 * @author Ay
 * @create 2019/08/17
 **/
public interface ProductRepository extends JpaRepository<AyProduct,Integer> {

}
```

在 com.example.demo.model 包下创建 AyUserKillProductRepository.java 类，负责用户秒杀商品记录的 CRUD（增删改查），具体代码如下：

```java
/**
 * 描述：用户秒杀商品记录 Repository
```

```
 * @author Ay
 * @create 2019/08/17
 **/
public interface AyUserKillProductRepository extends
 JpaRepository<AyUserKillProduct,Integer> {

}
```

21.2.4　Service 服务层的设计与开发

在 com.example.demo. service 包下创建商品接口类 ProductService.java，具体代码如下：

```
/**
 * 描述：商品接口
 * @author Ay
 * @create 2019/08/17
 **/
public interface ProductService {

    /**
     * 查询所有商品
     * @return
     */
    List<AyProduct> findAll();

    /**
     * 秒杀商品
     * @param productId 商品 id
     * @param userId 用户 id
     * @return
     */
    AyProduct killProduct(Integer productId, Integer userId);
}
```

在 com.example.demo. service 包下创建用户秒杀商品记录接口类 AyUserKillProductService.java，具体代码如下：

```
/**
 * 描述：用户秒杀商品记录接口
 * @author Ay
```

```
 * @create 2019/08/20
 **/
public interface AyUserKillProductService {

    /**
     * 保存用户秒杀商品记录
     * @param killProduct
     * @return
     */
    AyUserKillProduct save(AyUserKillProduct killProduct);
}
```

商品接口类 ProductService.java 和用户秒杀商品记录接口类 AyUserKillProductService.java 开发完成之后,接下来开发这两个接口对应的实现类。

在 com.example.demo.service.impl 包下创建 AyUserKillProductServiceImpl.java 类,该类用来实现 AyUserKillProductService 接口,同时实现 save 方法,具体代码如下:

```
/**
 * 描述:
 * @author Ay
 * @create 2019/08/20
 **/
@Service
public class AyUserKillProductServiceImpl implements AyUserKillProductService {
    //注入 ayUserKillProductRepository 类
    @Resource
    private AyUserKillProductRepository ayUserKillProductRepository;

    /**
     * 保存用户秒杀商品记录
     * @param killProduct
     * @return
     */
    @Override
    public AyUserKillProduct save(AyUserKillProduct killProduct) {
        return ayUserKillProductRepository.save(killProduct);
    }
}
```

在 com.example.demo.service.impl 包下创建 ProductServiceImpl.java 类，该类用于实现 ProductService.java 接口，并实现 findAll、killProduct 等方法，具体代码如下：

```java
/**
 * 描述：商品服务
 * @author Ay
 * @create 2019/08/17
 **/
@Service
public class ProductServiceImpl implements ProductService {

    @Resource
    private ProductRepository productRepository;

    @Resource
    private AyUserKillProductService ayUserKillProductService;
    //日志
    Logger logger = LoggerFactory.getLogger(ProductServiceImpl.class);

    /**
     * 查询所有商品
     * @return
     */
    @Override
    public List<AyProduct> findAll() {
        try{
            List<AyProduct> ayProducts = productRepository.findAll();
            return ayProducts;
        }catch (Exception e){
            logger.error("ProductServiceImpl.findAll error", e);
            return Collections.EMPTY_LIST;
        }
    }

    /**
     * 秒杀商品
     * @param productId 商品 id
     * @param userId 用户 id
     * @return
     */
```

```java
@Override
public AyProduct killProduct(Integer productId, Integer userId) {
    //查询商品
    AyProduct ayProduct = productRepository.findById(productId).get();
    //判断商品是否还有库存
    if(ayProduct.getNumber() < 0){
        return null;
    }
    //设置商品的库存：原库存数量 - 1
    ayProduct.setNumber(ayProduct.getNumber() - 1);
    //更新商品库存
    ayProduct = productRepository.save(ayProduct);
    //保存商品的秒杀记录
    AyUserKillProduct killProduct = new AyUserKillProduct();
    killProduct.setCreateTime(new Date());
    killProduct.setProductId(productId);
    killProduct.setUserId(userId);
    //设置秒杀状态
    killProduct.setState(KillStatus.SUCCESS.getCode());
    //保存秒杀记录详细信息
    ayUserKillProductService.save(killProduct);
    return ayProduct;
}
```

在 ProductServiceImpl.java 类中，注入 productRepository 和 ayUserKillProductService 类，之所以不直接注入 AyUserKillProductRepository 类，是因为 Service 类只能注入自己的 Repository 类，不能注入其他的 Repository 类。比如 ProductServiceImpl 注入自己的 ProductRepository 类，AyUserKillProductServiceImpl 类注入自己的 AyUserKillProductRepository 类。这一点要特别注意，在工作中非常容易犯错误。

21.2.5 Controller 控制层的设计与开发

在 com.example.demo.controller 包下创建控制层 ProductController.java，具体代码如下：

```
/**
 * 描述：
 * @author Ay
```

```java
 * @create 2019/08/17
 **/
@Controller
@RequestMapping("/products")
public class ProductController {

    @Resource
    private ProductService productService;

    /**
     * 查询所有的商品
     * @param model
     * @return
     */
    @RequestMapping("/all")
    public String findAll(Model model){
        List<AyProduct> products = productService.findAll();
        model.addAttribute("products", products);
        return "product_list";
    }

    /**
     * 秒杀商品
     * @param model
     * @param productId 商品id
     * @param userId 用户id
     * @return
     */
    @RequestMapping("/{id}/kill")
    public String killProduct(Model model,
                              @PathVariable("id") Integer productId,
                              @RequestParam("userId") Integer userId){
        AyProduct ayProduct = productService.killProduct(productId, userId);
        if(null != ayProduct){
            return "success";
        }
        return "fail";
    }
}
```

这里主要看方法的返回值，比如 findAll 方法最终返回"product_list"字符串，去请求 src\main\resources\templates 目录下的 product_list.html 页面，并返回给浏览器。

21.2.6　前端页面的设计与开发

后端代码基本开发完成，接下来简单开发前端页面。

首先，在 Spring Boot 中集成 thymeleaf 模版引擎，在 pom.xml 中添加 thymeleaf 依赖，具体代码如下：

```
<!-- thymeleaf 依赖 -->
<dependency>
    <groupId>org.springframework.boot</groupId>
    <artifactId>spring-boot-starter-thymeleaf</artifactId>
</dependency>
```

依赖添加完成后，在 application.properties 配置文件中添加 thymeleaf 相关的配置，具体代码如下：

```
#thymeleaf 配置
#模板的模式，支持如 HTML、XML、TEXT、JAVASCRIPT 等
spring.thymeleaf.mode=HTML5
#编码，可不用配置
spring.thymeleaf.encoding=UTF-8
#内容类别，可不用配置
spring.thymeleaf.servlet.content-type=text/html
#开发配置为 false，避免修改模板还要重启服务器
spring.thymeleaf.cache=false
#配置模板路径，默认就是 templates，可不用配置
#spring.thymeleaf.prefix=classpath:/templates/
```

在 src\main\resources\templates 目录下创建商品列表页面 product_list.html，具体代码如下：

```
<!DOCTYPE HTML>
<html xmlns:th="http://www.thymeleaf.org">
<head>
    <title>hello</title>
    <meta http-equiv="Content-Type" content="text/html; charset=UTF-8" />
    <!-- 引入 Bootstrap -->
```

```html
        <link href="https://maxcdn.bootstrapcdn.com/bootstrap/3.3.7/css/
bootstrap.min.css" rel="stylesheet">
    </head>
    <body>

    <!-- 页面显示部分-->
    <div class="container">
        <div class="panel panel-default">
            <div class="panel-heading text-center">
                <h1>秒杀活动</h1>
            </div>
            <div class="panel-body">
                <table class="table table-hover">
                    <thead>
                    <tr>
                        <td>图片</td>
                        <td>名称</td>
                        <td>库存</td>
                        <td>开始时间</td>
                        <td>结束时间</td>
                        <td>操作</td>
                    </tr>
                    </thead>

                    <tbody>
                    <tr th:each="product:${products}">
                        <td><img border="1px" width="100px" height="110px"
                            th:src="@{${product.productImg}}"/></td>
                        <td th:text="${product.name}"></td>
                        <td th:text="${product.number}"></td>
                        <td th:text="${product.startTime}"></td>
                        <td th:text="${product.endTime}"></td>
                        <td>
                            <!-- 发起秒杀请求,用户id先简单写死为 1 -->
                            <a class="btn btn-info" th:href="@{'/products/'
                            +${product.id}+ '/kill' + '?userId=1'}">秒杀</a>
                        </td>
                    </tr>
                    </tbody>
```

```html
            </table>
        </div>
    </div>
</div>

</body>
<!-- jQuery 文件。务必在 bootstrap.min.js 之前引入 -->
<script src="https://cdn.bootcss.com/jquery/2.1.1/jquery.min.js"></script>
<!-- 最新的 Bootstrap 核心 JavaScript 文件 -->
<script src="https://cdn.bootcss.com/bootstrap/3.3.7/js/bootstrap.min.js"></script>
</html>
```

在页面中，引入 jquery.min.js 和 bootstrap.min.js 文件，jQuery 是一个 JavaScript 库，可极大地简化 JavaScript 编程。Bootstrap 是 Twitter 公司推出的一个用于前端开发的开源工具包，它由 Twitter 的设计师 Mark Otto 和 Jacob Thornton 合作开发，是一个 CSS/HTML 框架，使用 Bootstrap，可以简单快速地让页面变得更漂亮，而 Bootstrap 依赖 jQuery，所以我们将 jQuery 也引入进来。更多 Bootstrap 和 jQuery 的知识，读者可到官方网站学习。

Bootstrap 官方网站地址：https://www.bootcss.com/
jQuery 官方网站地址：https://jquery.com/

在 src\main\resources\templates 目录下创建秒杀成功页面 success.html，具体代码如下：

```html
<!DOCTYPE HTML>
<html xmlns:th="http://www.thymeleaf.org">
<head>
    <title>hello</title>
    <meta http-equiv="Content-Type" content="text/html; charset=UTF-8" />
</head>
<body>
<div align="center">
    客官！！！  恭喜您，秒杀成功~~~
</div>
</body>
</html>
```

在 src\main\resources\templates 目录下创建秒杀失败页面 fail.html，具体代码如下：

```html
<!DOCTYPE HTML>
<html xmlns:th="http://www.thymeleaf.org">
<head>
    <title>hello</title>
    <meta http-equiv="Content-Type" content="text/html; charset=UTF-8" />
</head>
<body>
<div align="center">
    不要灰心，继续加油~~~
</div>
</body>

</html>
```

在 src\main\resources\static 目录下添加商品图片 spring-5.jpg、spring-boot.jpg、spring-mvc.jpg，商品图片信息如图 21-2 所示。

图 21-2　商品图片

21.2.7　代码测试

所有代码开发完成后，运行入口类 DemoApplication 的 main 方法启动项目，在浏览器中输入访问地址：http://localhost:8080/products/all，便可以查询到商品列表，具体页面如图 21-3 所示。

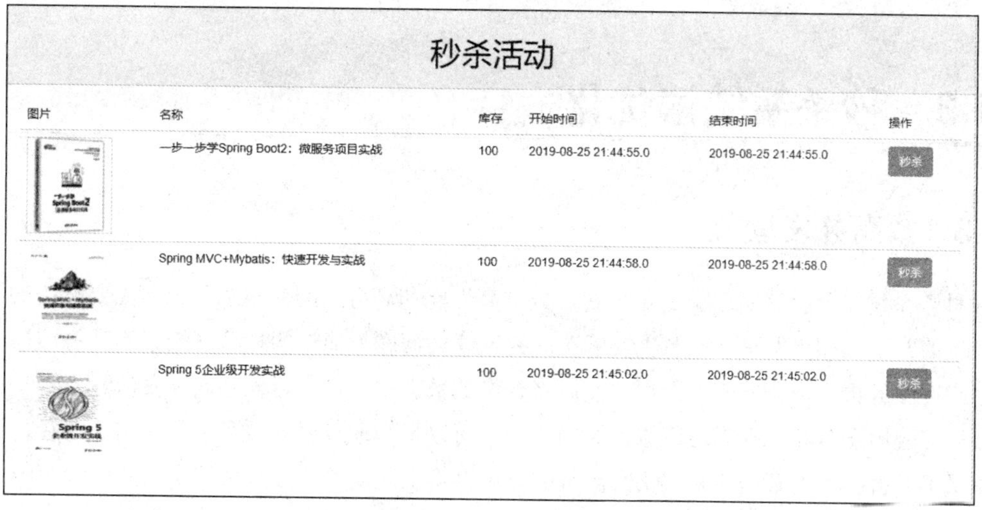

图 21-3 秒杀活动商品列表页面

单击秒杀按钮，前端页面发起 URL 请求：http://localhost:8080/products/1/kill?userId=1，调用后端控制层 ProductController 的方法 killProduct，进行商品秒杀。

21.2.8 总结

21.2 节主要使用简单的方式搭建了一个可用的秒杀系统，其系统架构非常简单，如图 21-4 所示。

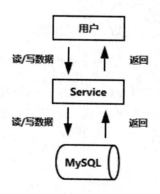

图 21-4 简单的秒杀系统架构

用户发起查询或者秒杀请求，服务端接收到请求后，向数据库发起读或者写的请求，然后将结果返回给用户。这样的简单系统架构，看似合理，但是当有大量的并发读和并发写操作时，高并发流量会使系统服务或者数据库出现故障或者宕机。因此，我们需要继续优化该秒杀系统。

21.3 秒杀系统读优化

21.3.1 高并发读优化

秒杀系统本质上就是一个满足大并发数、高性能和高可用的分布式系统，主要解决的问题是大量的并发读和并发写，本节主要解决高并发读的问题。简单的秒杀系统架构其每次请求先通过应用服务再到数据库查询数据，然后将查询的数据由下及上返回。高并发流量及过长的请求路径对应用服务和数据库会造成巨大的压力，可以考虑通过引入缓存，缩短请求路径，让请求流量不要直接查询数据库来解决大流量的。可以将上述简单的秒杀系统架构演化为如图 21-5 所示的系统架构。

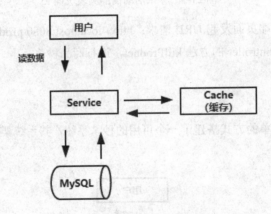

图 21-5　引入缓存的秒杀系统架构

用户查询秒杀商品列表时，先到缓存查询数据，如果缓存有用户需要的数据，直接返回给用户；否则，查询数据库，将数据存入到缓存中，最后将数据返回给用户。因为秒杀商品属于不经常修改的数据，所以非常适合存放在缓存中。

21.3.2 使用 Redis 缓存

目前市面上开源的缓存产品很多，我们就以目前企业中流行的 Redis 缓存为例，将其集成到 Spring Boot 项目中。首先，在 pom.xml 文件中添加 Redis 缓存的相关依赖，具体代码如下：

```
<!-- redis start -->
<dependency>
```

```
    <groupId>org.springframework.boot</groupId>
    <artifactId>spring-boot-starter-data-redis</artifactId>
</dependency>
```

在 application.properties 文件中添加 Redis 配置，具体代码如下：

```
### Redis 缓存配置
### 默认 Redis 数据库为 db0
spring.redis.database=0
### 服务器地址，默认为 localhost
spring.redis.host=localhost
### 链接端口，默认为 6379
spring.redis.port=6379
### Redis 密码默认为空
spring.redis.password=
```

在 ProductService.java 类中添加接口 findAllCache，该方法用于查询商品列表数据，只是带有缓存功能，具体代码如下：

```
/**
 * 查询所有商品
 * @return
 */
Collection<AyProduct> findAllCache();
```

在 ProductServiceImpl.java 类中实现 findAllCache 接口，具体代码如下：

```
//注入 redisTemplate 对象
    @Resource
    private RedisTemplate redisTemplate;
    //定义缓存 key
    private static final String KILL_PRODUCT_LIST = "kill_product_list";

    /**
     * 查询商品数据（带缓存）
     * @return
     */
    @Override
    public Collection<AyProduct> findAllCache() {
        try{
```

```java
        //从缓存中查询商品数据
        Map<Integer, AyProduct> productMap =
                redisTemplate.opsForHash().entries(KILL_PRODUCT_LIST);
        Collection<AyProduct> ayProducts = null;
        //如果缓存中查询不到商品数据
        if(CollectionUtils.isEmpty(productMap)){
            //从数据库中查询商品数据
            ayProducts = productRepository.findAll();
            //将商品list转换为商品map
            productMap = convertToMap(ayProducts);
            //将商品数据保存到缓存中
            redisTemplate.opsForHash().putAll(KILL_PRODUCT_LIST, productMap);
            //设置缓存数据的过期时间,这里设置为10秒,具体时间需要结合业务需求而定
            //如果商品数据变化少,过期时间可以设置长一点;反之,过期时间可以设置短一点
            redisTemplate.expire(KILL_PRODUCT_LIST,10000 ,
TimeUnit.MILLISECONDS);
            return ayProducts;
        }
        ayProducts = productMap.values();
        return ayProducts;
    }catch (Exception e){
        logger.error("ProductServiceImpl.findAllCache error", e);
        return Collections.EMPTY_LIST;
    }
}

/**
 * list转换为map
 * @param ayProducts
 * @return
 */
private Map<Integer, AyProduct> convertToMap(Collection<AyProduct> ayProducts){
    if(CollectionUtils.isEmpty(ayProducts)){
        return Collections.EMPTY_MAP;
    }
    Map<Integer, AyProduct> productMap = new HashMap<>(ayProducts.size());
    for(AyProduct product: ayProducts){
```

```
        productMap.put(product.getId(), product);
    }
    return productMap;
}
```

redisTemplate.opsForHash().entries (K key)：通过 entries(H key)方法获取变量中的键值对。

在 ProductController.java 中添加方法 findAllCache 方法，具体代码如下：

```
/**
 * 查询所有的商品（缓存）
 * @param model
 * @return
 */
@RequestMapping("/all/cache")
public String findAllCache(Model model){
    Collection<AyProduct> products = productService.findAllCache();
    model.addAttribute("products", products);
    return "product_list";
}
```

代码开发完成之后，停止并重新启动项目（运行入口类的 main 方法），在浏览器中输入访问地址：http://localhost:8080/products/all/cache，查询秒杀系统的商品列表页，可以发现第一次查询，后端服务需要从数据库中查询商品列表，并将数据保存到缓存中，再次刷新商品列表页，数据会直接从缓存中获取，返回给用户。

单击秒杀按钮，秒杀成功后，重新回到秒杀商品列表，发现商品的库存数量没有减少，原因是缓存中的数据还是旧的商品数据，缓存中的商品数据在没有过期的这段时间内，用户会查询到"脏的数据"，虽然不影响秒杀功能，但是我们还是需要修正该问题。

修复该问题的方法很简单，只需要在 ProductServiceImpl.java 类中的 killProduct 方法后面添加如下代码即可：

```
//商品秒杀成功后，更新缓存中商品库存数量
redisTemplate.opsForHash().put(KILL_PRODUCT_LIST, killProduct.getProductId(), ayProduct);
```

商品秒杀成功后，更新缓存中的商品库存数量。redisTemplate.opsForHash().put (H key, HK hashKey, HV value)用于更新缓存数据，hashKey 为 productId，value 为商品实体。

21.4 流量削峰

21.4.1 流量削峰的原因

由于秒杀请求在短时间内高度集中于某一特定的时间点，会导致一个特别高的流量峰值，它对资源的消耗是瞬时的。一台服务器处理资源的能力是固定的，如果出现流量峰值的话，很容易造成系统的瓶颈，况且最终能够秒杀到商品的请求是固定的，比如 1 万个秒杀请求到最后真正能成功的请求可能只有 1000 个。因此，我们需要设计一些原则，让并发的请求更加平缓有序地进行，这也是为什么需要流量削峰的原因。

流量削峰，最容易想到的就是消息队列。可使用消息队列来缓冲瞬时流量，将同步请求转换为异步请求。

21.4.2 集成 ActiveMQ

在 Spring Boot 中集成 ActiveMQ，首先需要在 pom.xml 文件中引入所需的依赖，具体代码如下：

```xml
<!-- activemq start -->
<dependency>
    <groupId>org.springframework.boot</groupId>
    <artifactId>spring-boot-starter-activemq</artifactId>
</dependency>
```

依赖添加完成之后，在 application.properties 配置文件中添加 ActiveMQ 配置，具体代码如下：

```
spring.activemq.broker-url=tcp://localhost:61616
spring.activemq.in-memory=true
spring.activemq.pool.enabled=false
spring.activemq.packages.trust-all=true
```

- spring.activemq.packages.trust-all：由于 ObjectMessage 的使用机制是不安全的，因此 ActiveMQ 自 5.12.2 和 5.13.0 版本之后，强制 Consumer 端声明一份可信任的包列表，只有当 ObjectMessage 中的 Object 在可信任包内，才能被提取出来。该配置表示信任所有的包。

关于如何安装 ActiveMQ，请读者参考第 11 章的内容，这里不再重复赘述。

21.4.3 生产者开发

在 \src\main\java 目录下创建 com.example.demo.producer 包，在该包下创建 ProductController.java 类，具体代码如下：

```java
/**
 * 生产者
 * @author Ay
 * @create 2019/08/31
 **/
@Service
public class AyProductKillProducer {

    //日志
    Logger logger = LoggerFactory.getLogger(ProductServiceImpl.class);

    @Resource
    private JmsMessagingTemplate jmsMessagingTemplate;

    /**
     * 描述：发送消息
     * @param destination 目标地址
     * @param killProduct 描述商品
     */
    public void sendMessage(Destination destination, final AyUserKillProduct killProduct) {
        logger.info("AyProductKillProducer sendMessage , killProduct is" + killProduct);
        jmsMessagingTemplate.convertAndSend(destination, killProduct);
    }
}
```

上述代码中，注入了 JmsMessagingTemplate，也可以注入 JmsTemplate。JmsMessagingTemplate 是对 JmsTemplate 进行了封装。参数 destination 是发送到的队列，message 是待发送的消息。

21.4.4 消费者开发

在 \src\main\java 目录下创建 com.example.demo.consumer 包，在该包下创建 AyProductKillConsumer.java 类，具体代码如下：

```java
/**
 * 消费者
 * @author Ay
 * @date   2017/11/30
 */
@Component
public class AyProductKillConsumer {

    //日志
    Logger logger = LoggerFactory.getLogger(ProductServiceImpl.class);

    @Resource
    private AyUserKillProductService ayUserKillProductService;

    /**
     * 消费消息
     * @param killProduct
     */
    @JmsListener(destination = "ay.queue.asyn.save")
    public void receiveQueue(AyUserKillProduct killProduct){
        //保存秒杀商品数据
        ayUserKillProductService.save(killProduct);
        //记录日志
        logger.info("ayUserKillProductService save, and killProduct: " + killProduct);
    }
}
```

- @JmsListener：使用 JmsListener 配置消费者监听的队列 ay.queue.asyn.save，其中 AyUserKillProduct 是秒杀商品。

最后，修改 ProductServiceImpl.java 类中的 killProduct 方法，具体代码如下：

```java
//注入 AyProductKillProducer 类
@Resource
private AyProductKillProducer ayProductKillProducer;

//定义消息队列
private static Destination destination = new ActiveMQQueue("ay.queue.asyn.save");

/**
 * 秒杀商品(引入MQ)
 * @param productId 商品 id
 * @param userId 用户 id
 * @return
 */
@Override
public AyProduct killProduct(Integer productId, Integer userId) {
    //查询商品
    AyProduct ayProduct = productRepository.findById(productId).get();
    //判断商品是否还有库存
    if(ayProduct.getNumber() < 0){
        return null;
    }
    //设置商品的库存：原库存数量 - 1
    ayProduct.setNumber(ayProduct.getNumber() - 1);
    //更新商品库存
    ayProduct = productRepository.save(ayProduct);
    //保存商品的秒杀记录
    AyUserKillProduct killProduct = new AyUserKillProduct();
    killProduct.setCreateTime(new Date());
    killProduct.setProductId(productId);
    killProduct.setUserId(userId);
    //设置秒杀状态
    killProduct.setState(KillStatus.SUCCESS.getCode());
    //保存秒杀记录详细信息
    //ayUserKillProductService.save(killProduct);
    //异步保存商品的秒杀记录（最重要的一句）
```

```
        ayProductKillProducer.sendMessage(destination, killProduct);
        //商品秒杀成功后,更新缓存中商品库存数量
        redisTemplate.opsForHash().put(KILL_PRODUCT_LIST,
killProduct.getProductId(), ayProduct);
        return ayProduct;
    }
```

上述代码中,将同步保存秒杀商品记录修改为异步保存,一方面可以提高请求的响应速度,提高用户的使用体验;另一方面减少了流量高峰对数据库的压力。当然,我们只是修改一部分同步保存方法,其他的方法读者可以继续优化。例如,可以把更新商品库存修改为异步(productRepository.save(ayProduct)),当异步更新完商品库存时,通知用户秒杀成功。

21.5 业务优化

21.5.1 答题/验证码

使用 12306 买过火车票的读者一定知道,用户抢购火车票时,需要进行答题抢票。例如图 21-6 所示的题目。

答题抢票的主要目的是,除了防止黄牛和抢票软件,区分人工和机器外,深层次的原因是增加用户购买的难度,延缓请求。抢票难度增大后,用户下单的时间会增加,从之前的 1~2 秒之内延长到 2 秒之后,这个时间的延缓对于后端服务处理并发非常重要,会大大减少压力。

图 21-6　12306 答题抢票

答题形式除了图 12-6 的方式,还可以有其他的方式,例如:

(1)计算一个简单的数学题。
(2)输入验证码。
(3)输入一串中文。
(4)回答一个调皮的问题。

21.5.2 分时分段

细心的用户可能会注意到,12306 放票不是一次性放完,而是分成几个时间段来放的。例如,在原来 8:00 至 18:00（除 14:00 外）每整点放票的基础上,增加 9:30、10:30、12:30、13:30、14:00、14:30 6 个放票时间点,每隔半小时放出一批,将流量摊匀。

21.5.3 禁用秒杀按钮

用户在单击秒杀按钮后,如果后端服务压力大,系统响应时间长,用户基本都会再次单击,一直单击秒杀按钮。每单击一次秒杀按钮,都会向后端服务请求一次。这样请求越来越多,系统压力越来越大,最后会造成任何用户都没法秒杀成功。

一种很简单的解决方法是,用户单击"秒杀"按钮后,按钮为灰色,表示禁止用户继续提交秒杀请求。同时限制用户在 x 秒内只能提交一次请求（根据具体的业务,设置合理的 x 时间,比如 5 秒、10 秒等）,如图 21-7 所示。

图 21-7　禁用秒杀按钮

21.6　降级、限流、拒绝服务

21.6.1 降级

在业务高峰时,为了保证服务的高可用,往往需要服务或者页面有策略地不处理或换一种简单的方式处理,从而释放服务器资源以保证核心交易正常运作或高效运作。这种技术在分布式微服务架构中称为服务降级。

例如在线购物系统,整个购买流程是重点业务,比如支付功能在流量高峰时,为了保证购买流程的正常运行,往往会关闭一些不太重要的业务（如广告业务等）。

降级方案可以这样设计:当秒杀流量达到 10 万/s 时,把成交记录的获取从展示 20 条降级到只展示 10 条。"从 20 改到 5"这个操作由一个开关来实现,也就是设置一个能够从开关系统动态获取的系统参数。

21.6.2 限流

限流是指通过对某一时间窗口内的请求数进行限制,保持系统的可用性和稳定性,防止因流量暴增而导致系统运行缓慢或宕机,限流的根本目的就是为了保障服务的高可用。当系统容量达到瓶颈时,我们需要通过限制一部分流量来保护系统,并做到既可以人工执行开关,也支持自动化保护的措施。限流是比降级更极端的一种保护措施。

限流既可以是在客户端限流,也可以是在服务端限流。限流的实现方式既要支持URL以及方法级别的限流,也要支持基于QPS(每秒查询率)和线程的限流。例如,我们的系统最高支持1万QPS时,可以设置8000 QPS来进行限流保护。

21.6.3 拒绝服务

如果系统流量实在太大,严重超出了系统的负载,系统可以直接拒绝所有的请求。比如,HTTP请求直接返回503错误码。拒绝服务是一种不得已的方案,同时也是最暴力最有效的系统保护方法。

系统虽然在过载时无法提供服务,但是仍然可以运作,当负载下降时又很容易恢复,所以每个系统和每个环节都应该设置这个方案,以对系统做最坏情况下的保护。

21.7 避免单点

所谓单点,即单机部署。单点意味着没有备份,风险不可控,一旦单点出问题,整个服务将不可用。

单点并不单指服务,也包括服务依赖的资源,比如数据库、缓存、消息中间件等。如图21-8所示是一个可变单点的集群服务。

一个系统不仅仅是单个应用为用户提供服务,而是采用集群的方法统一对外提供服务。如果服务A宕机,服务B、服务C及服务D仍可以继续提供服务,并不会影响整个系统。除了应用服务需要避免单点外,数据库以及缓存都可以通过改造(比如数据库主从设计、缓存主从设计等)来避免单点服务。

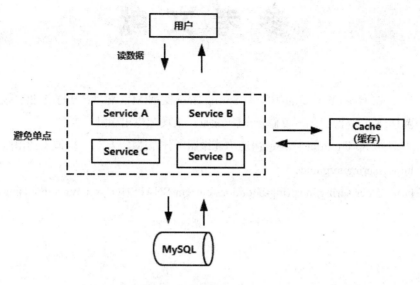

图 21-8　服务集群，避免单点

21.8　总结

秒杀系统是学习"三高"（高性能、高并发、高可用）的一个非常好的例子，具有读多写少等特性。为了保证系统的高可用，使用避免单点措施（应用服务、数据库、缓存等）可保证服务的稳定性；对于高并发读，引入 Redis 缓存可避免流量直接穿透到数据库，同时，引入消息中间件则可对流量进行削峰。除了技术上的设计，业务方法上使用答题/验证码、分时分段以及禁用秒杀按钮等措施，可将请求流量尽量拦截在上游。最后，在最坏的情况下，使用系统降级、限流、拒绝服务等错误，可起到保护系统以防宕机的作用。为了让读者能轻松学习，本书仅提供了一个简单的例子，真实的秒杀系统架构设计更复杂，业务逻辑更烦琐，读者可在此基础上进一步完善。

参 考 文 献

[1] 汪云飞. Java EE 开发的颠覆者 Spring Boot 实战[M]. 北京：电子工业出版社，2016.
[2] 郝佳. Spring 源代码深度解析[M]. 北京：人民邮电出版社，2013.
[3] 王富强. Spring Boot 揭秘：快速构建微服务体系[M]. 北京：机械工业出版社，2016.
[4] https://spring.io/guides
[5] https://docs.spring.io/spring-boot/docs/current-SNAPSHOT/reference/htmlsingle/